Ethnographies of Science Education

Researchers who conduct ethnography in science education tend to have a deep commitment for transforming science to improve the lives of people in underserved communities. This edited volume explores how contemporary ethnographers in science education bring to light the local production of scientific knowledge and the ways it is implicated in larger social and political struggles. Ethnographies in science education contribute to understanding the experiences of linguistically, racially, and economically diverse populations who have been historically excluded from participation in science. An anthropological approach has also been instrumental in explicating the situated practices by which students enact science in the classroom and in their lives beyond schools.

This edited volume brings together ethnographers in science education to contribute a global perspective on science teaching and learning in school and university classrooms, at home, and in afterschool programs. Included are examples of ethnography in science education from the UK, Argentina, Canada, and the USA in which contributors point to promising directions for theorizing the 'culture' of science education as we undertake educational reform. The authors in this volume argue that ethnography is not only a valid approach for the study of science education, but also, they contend that it is essential to the development of more equitable practices for instruction and learning.

This book was originally published as a special issue of *Ethnography and Education.*

Carol B. Brandt is an Assistant Professor of Science Education in the Department of Teaching and Learning at Temple University in Philadelphia, Pennsylvania, USA. As an educational anthropologist, her research explores the sociocultural dimensions of learning science in informal contexts beyond the school classroom. She studies the ways that language structures participation as youth and adults move between home, community, science centers, and school.

Heidi Carlone is an Associate Professor of Science Education at the University of North Carolina at Greensboro, USA. Professor Carlone's research uses theoretical lenses from anthropology of education, sociocultural theories, and cultural studies to understand the local/global productions of 'science' and 'science person' in K-16 in-school and out-of-school settings and the implications of those meanings for those historically shut out of science and for science education reform.

Ethnographies of Science Education

Situated Practices of Science Learning for Social/Political Transformation

Edited by
Carol B. Brandt and Heidi Carlone

Routledge
Taylor & Francis Group
LONDON AND NEW YORK

First published 2014
by Routledge
4 Park Square, Milton Park, Abingdon, Oxon OX14 4RN
605 Third Avenue, New York, NY 10017

Routledge is an imprint of the Taylor & Francis Group, an informa business

British Library Cataloguing in Publication Data
A catalogue record for this book is available from the British Library

ISBN 13: 978-1-138-02062-7 (hbk)

Typeset in Times New Roman
by Taylor & Francis Books

Publisher's Note
The publisher accepts responsibility for any inconsistencies that may have arisen during the conversion of this book from journal articles to book chapters, namely the possible inclusion of journal terminology.

Contents

Citation Information

The chapters in this book were originally published in *Ethnography and Education*, volume 7, issue 2 (June 2012). When citing this material, please use the original page numbering for each article, as follows:

Chapter 1
Guest Introduction: Ethnographies of science education: situated practices of science learning for social/political transformation
Carol B. Brandt and Heidi Carlone
Ethnography and Education, volume 7, issue 2 (June 2012) pp. 143–150

Chapter 2
Unpacking 'culture' in cultural studies of science education: cultural difference versus cultural production
Heidi Carlone and Angela Johnson
Ethnography and Education, volume 7, issue 2 (June 2012) pp. 151–174

Chapter 3
From science as 'content' to science as 'interpretive key': experiences and reflections from a science course in teacher education
Laura Colucci-Gray and Christine Fraser
Ethnography and Education, volume 7, issue 2 (June 2012) pp. 175–196

Chapter 4
Evolution education in policy and practice: an ethnographic perspective
David E. Long
Ethnography and Education, volume 7, issue 2 (June 2012) pp. 197–212

Chapter 5
Science teaching: a dilemmatic approach
Anna Traianou
Ethnography and Education, volume 7, issue 2 (June 2012) pp. 213–226

Chapter 6
Feeding pigs and looking for güembé*: the local production of knowledge about the natural world of peasant and indigenous children in San Ignacio*
Ana Padawer
Ethnography and Education, volume 7, issue 2 (June 2012) pp. 227–246

Chapter 7

Collaborative imaginaries and multi-sited ethnography: space-time dimensions of engagement in an afterschool science programme for girls
Jrene Rahm
Ethnography and Education, volume 7, issue 2 (June 2012) pp. 247–264

Chapter 8

The role of borders in environmental education: positioning, power and marginality
Carrie T. Tzou and Philip Bell
Ethnography and Education, volume 7, issue 2 (June 2012) pp. 265–282

Please direct any queries you may have about the citations to clsuk.permissions@cengage.com

Notes on Contributors

Philip Bell, Learning Sciences and Human Development, College of Education, University of Washington Seattle, WA, USA

Carol B. Brandt, Department of Teaching and Learning, Temple University, Philadelphia, PA, USA

Heidi Carlone, Department of Teacher Education and Higher Education, University of North Carolina at Greensboro, Greensboro, NC, USA

Laura Colucci-Gray, School of Education, University of Aberdeen, Aberdeen, UK

Christine Fraser, School of Education, University of Aberdeen, Aberdeen, UK

Angela Johnson, Department of Educational Studies, St. Mary's College of Maryland, St. Mary's City, MD, USA

David E. Long, Centre for Reconstructing Education in Science and Technology, George Mason University, Fairfax, VA, USA

Ana Padawer, Instituto de Ciencias Antropológicas de la Universidad de Buenos Aires y Consejo Nacional de Investigaciones Científicas y Tecnológicas (CONICET), Buenos Aires, Argentina

Jrene Rahm, Département de psychopédagogie et d'andragogie, Université de Montréal, Montréal, Canada

Anna Traianou, Department of Educational Studies, Goldsmiths, University of London, New Cross, London, UK

Carrie T. Tzou, Education Program, University of Washington Bothell, Bothell, WA, USA

INTRODUCTION

Ethnographies of science education: situated practices of science learning for social/political transformation

Transforming science education has been cited as a global imperative in terms of: producing technological innovation to maintain economic security (Bybee and Fuchs 2006; Tytler 2007); creating critical consumers of scientific knowledge (Osborne and Dillon 2008) and fostering a more environmentally sustainable and equitable world (Calabrese Barton 2001; Carter 2008). In light of these agendas, viewing science as a cultural process has significantly contributed to our understanding of the interplay between local micro-level contexts and macro-level political influences in the science classroom. Moreover, ethnography has chronicled the experiences of ethnically and linguistically diverse populations who have been historically excluded from participation in science. Cultural studies of science education speak directly to issues of economics, sustainability and inclusion but also address theoretical and empirical gaps in our understanding of science education and its context: 'What precisely is the nature of science, of nature, of culture, and of the relationship among them?' (Weinstein 1998, 486).

Researchers who conduct ethnography in science education tend to have a deep commitment for transforming science to become an agentic tool, one that improves the lives of people in underserved communities (Hammond and Brandt 2004). In this light, identity and agency – the human capacity for making choices and the ability to act upon these intentions – is viewed as being both important in terms of learning science, as well as understanding social change in schools and the broader society. Yet, by taking up this stance, the ethnographer in science education is often at odds with the very practices that distinguish the sciences as a process of inquiry separate from other disciplines. Ethnographers of science education have opened up the science classroom to describe cultural practices surrounding the teaching and learning of science in the same way that sociologists have studied the construction of knowledge in the science laboratory (Collins 1982; Latour and Woolgar 1979/1986) and the socialisation of scientists-in-the-making (Knorr-Cetina and Mulkay 1983). Through their research, ethnographers of science education challenge the 'culture of no culture' (Subramaniam and Wyer 1998; Traweek 1988) and the prevalent myth of objectivism in science. As ethnographers examine the ways science is given meaning in schools, they ask: *What* is science education? *Whose* purposes does it support?

This special journal issue explores how contemporary ethnographers in science education study the local production of scientific knowledge and how this meaning-making is implicated in larger social and political struggles. The articles in this issue have a two-fold purpose. First, these articles offer examples of the socially and politically situated practices of science learning (in- and out-of-school contexts). Second, these articles highlight the tensions in critically examining science as a social practice while re-envisioning science education and science education policy. In this issue, we reconsider science education and the ways in which ethnographic research can contribute to its transformation. We argue that a cultural perspective of science education is essential for the design of more effective interventions – ones that expose, confront and seek to alter persistent structural inequities in schooling, while also addressing pressing global concerns of economic and environmental sustainability, health and social justice.

The ethnographic turn in science education

Ethnographies of science education are a relatively recent addition to educational research. As Delamont (1989) noted in her review of sociological approaches to science education, prior to the 1980s, sociological accounts of science classrooms in the UK were completely missing – and similarly, ethnography of science education in North American was non-existent. Prior to the 1980s, the emphasis in science education research was to emulate experimental studies in the scientific journals such as *Nature* or *Science* (Joslin et al. 2008). With the rise of social constructivism in science education research the late 1980s (Driver and Oldham 1986), research in education moved from experimental studies focused on individual cognition to include studies of the context of science classrooms and the practices of teaching and learning science. In their international review of ethnographic studies in science education, Hammond and Brandt (2004) noted that few researchers in science education position themselves as anthropologists and similarly, the use of the term 'ethnography' or 'ethnographic' varies widely. Some researchers claim their work as ethnographic but are little more than interview studies or descriptive qualitative research in a naturalistic setting. Other qualitative studies provide microanalyses of social interactions of learning but avoid taking up a position on what constitutes 'culture' and the ways science education is reproduced in schools.

The ethnographic turn in science education did not emerge directly out of anthropology or sociology, nor was it influenced by the ways that early anthropologists or sociologists thought about culture. Instead, the work of Vygotsky (1978, 1986) and other Russian psychologists (Leont'ev 1978) in the early twentieth century played a major theoretical role in a sociocultural approach to learning and signalled a radical departure from Piaget's (1955) model of cognition. The turn towards understanding human practices of learning emerged from an interest on the context of socially situated activity (Lave 1993, 1997) and the ways that knowledge is socially constructed (Bruner 1996; Cole 1971). Naturalistic studies of classroom practice (Brown 1992; Brown and Campione 1994) developed from the field of cultural psychology whose researchers began to study the ways that social interactions and the role of language function in human learning.

When naturalistic studies of classrooms were introduced in academic journals during the late 1970s, controversy brewed in science education (Joslin et al. 2008) surrounding the appropriateness of qualitative methods for research and whether naturalistic studies could even be considered 'research' at all. Despite the dominance of positivism in educational research, early ethnographers in education were inspired by the work of Erickson (1986) to undertake detailed descriptions of classroom learning. Researchers like Ken Tobin (Siry 2009) argued for a departure from

Piagetian theory to view science – and science education – as social and cultural process.

We emphasise the origins of an ethnographic turn in science education to underscore the ways that the individual is theorised in relation to larger cultural processes and institutions. Reform in science education (as in education in general) has consistently focused on measuring the individual – the role of the teacher or assessing the motivation and success of the student – without considering a more systemic approach to examining cultural, institutional or global contexts of educational resources and knowledge. Ethnography is one way to broaden this lens to incorporate those aspects of human practice formerly viewed as peripheral, deemed unimportant or judged as ancillary to science education.

Elusive venues for the publication of ethnography in science education

Science education journals in the early 1980s began opening the door to the publication of a wide variety of qualitative studies and with these, the very first ethnographies of science education.[1] Other academic journals quickly followed suit and several top-ranked journals in science education currently publish ethnographic works alongside experimental studies. Even though international journals in science education now provide opportunities for publication of qualitative research, ethnographic studies have yet to be published in these venues. Despite a palpable shift in research in science education in North America, Australia and the UK, a 'scientific' approach to conducting educational research still dominates in South American, Asian, African, Eastern European and European science education journals – providing few locations for the publication of ethnographic studies. Our review of ethnography in science education for this overview indicates that those researchers conducting science education in these countries either publish in North American science education journals or in the fields of anthropology and sociology.

The 'gatekeeping' effect of academic journals on ethnography in science education is undeniable, yet other factors have also influenced the rarity of ethnographic studies in science education. Despite a general increase in qualitative research in education, 'scientifically based research' has been favoured for grant funding and designated as superior through national agendas and government policies (Berliner 2002; Eisenhart 2006; Phillips 2002, 2005). In this paradigm, experimental studies, hypothesis testing and control groups are favoured, while ethnography is deemed a 'soft' approach lacking validity or generalisability. Similarly, many graduate programmes in science education now include instruction in qualitative approaches (interviewing, participant observation and the collection/analysis of cultural artefacts) especially if taught as complementary to quantitative measures and sanctioned as part of 'mixed methods'. Yet, ethnography in graduate education as both methodology *and* ontology continues to be uncommon.

In sum, as education research continues to open up dialogues around 'ecological' frameworks of learning and development (Barab and Roth 2006; Barron 2006; Bronfenbrenner 1979; Lee 2008), we see an opportunity to promote ethnography as a means of advancing more equitable practices in science education. In addition, we wish to encourage science education to look beyond the individual to consider how meaning is shared widely, that is – how science 'culture' is expressed and implicated across sites of learning. Too often, sociocultural studies in science education end at the classroom walls and the interplay between micro-level practices and larger macro-level institutions go unacknowledged. Also, with a focus on the individual (or individual cognition), group/institutional norms tied to discriminatory and exclusionary practices often remain unchallenged and unexamined.

The organisation of this special issue

This special issue brings together ethnographers in science education to contribute a global perspective on science teaching and learning in school and university classrooms, at home and after school programmes. The settings and actors are diverse and constitute 'science', defined broadly. This issue includes examples of ethnography in science education from the UK, Argentina, Canada and USA. Four of the contributions are focused on science education in public schooling, while the other three examine science learning in out-of-school contexts. Three cross-cutting themes dominate the articles in this issue: (1) ethnography as an approach to articulate the tensions between institutional norms of 'scientific knowledge' and local meaning-making that occurs in the lives of students and teachers; (2) ethnography as an analytical tool for teacher self-reflection and transformative action and (3) ethnography as a strategy to map the ways that youth position themselves or are positioned through time and space as they participate in science education.

Carlone and Johnson's article leads this issue by exploring the various ways that culture is theorised in science education. The authors take up three anthropological approaches common in science education (funds of knowledge, third space/hybridity and practice theory) to debate the advantages and limitations of each framework. Carlone and Johnson raise an important point, whereby researchers attend closely to the ways in which cultural repertoires (or particular cultural elements) are elevated and are unintentionally essentialised in an attempt to explain inequities in schooling. The authors point out the potential hazard in which theoretical underpinnings fail to account for the ways science as a culture reproduces itself, and how science is connected to larger systems of power. By examining 4 years of ethnographic data in American public school science classrooms, Carlone and Johnson illustrate the need for a theoretical lens that can accommodate nuances and subtle differences in local practices and the socio-historical shaping of local meaning over time. Using the case of Julio, a Mexican immigrant student, they argue that the strength of practice theory lies in its ability to view meaning-making among individuals and groups, as being shaped *and* shaping larger macro-level structures.

In their ethnography of a primary teacher education programme in the UK, Colucci-Gray and Fraser adopt a framework based on a heightened awareness of language and one's own experience as an 'interpretative key' in a science education class. Taught in a way that emphasised students' co-construction of knowledge and science as cultural process, this unfamiliar strategy created unexpected tensions for the students and their instructors, especially when students demanded content and a more rigid presentation of science. Colucci-Gray and Fraser found those students who had experience in cultural 'border-crossing' at the university (in this case, Spanish exchange students) were better able to adapt to this innovative curriculum. These authors ask important questions concerning the design of elementary teacher education whereby potential teachers (and consequently their own students) are empowered towards greater democratic participation in science. Articulating these tensions was a decisive step towards the redesign of their course.

Likewise, Long's ethnography exposed tensions among the students and faculty surrounding the presentation of evolution in college-level biology courses. In his ethnography of undergraduates and faculty at regional university in the religiously conservative Midwestern USA, Long explored the ways students who identify as Creationists struggled with describing conflicted feelings about evolution while participating in the science course. Faculty, however, glossed-over and denied students a discursive space in which they could examine the nature of science in relation to their religious beliefs. Long views ethnography as a means for articulating what seem to be incommensurable beliefs about science and schooling, but also as an avenue for informing educational policy that promises more democratic participation.

Traianou's ethnography of a primary schoolteacher in the UK examined the use of a 'dilemmatic' approach for one's self-examination of science teaching. By identifying the contradictory patterns of schooling, teachers can identify tools for investigating practice and transforming it. In the context of high stakes testing, the teacher in Traianou's study confronted several dilemmas during the teaching of physical forces; these dilemmas highlighted aspects of her practice that grappled with societal norms of what constitutes 'scientific' knowledge and logical reasoning as opposed to her students' everyday meaning-making. The teacher in Traianou's study used the language of dilemmas to explore the competing pressures under which teachers operate and their thinking in the decision-making process. Traianou argues that using a dilemmatic approach in teacher education can facilitate an examination of one's own practice, an essential element of being a *bricoleur* who is engaged in boundary work, which requires a new trans-disciplinary consciousness (Kincheloe 2001).

Padawer's ethnography of children's knowledge about the natural world in the *Mbyá-Guaraní* communities of Argentina typifies much of rural South America where indigenous communities find themselves swept up in rapid economic and environmental change. As a backdrop to her ethnography, Padawer described the colonial past in rural Argentina whereby peasant squatters have encroached on indigenous communities through their participation in agriculture and an economy based on the extraction of forest products. As might be expected, the knowledge base of peasant and indigenous youth in the same school was the result of economic practices they shared with adults outside their homes. However, as recent global flows have shifted competition – economic roles and land ownership have been altered as well, impacting the ways that youth learn about the natural world and local ecology. By mapping local knowledge among youth and the spaces in which they construct these understandings, Padawer advocates that schools take advantage of children's everyday experiences to teach science education in schools and to challenge naive textbook representations of indigenous and peasant land-use practices.

In her ethnography of immigrant girls in an afterschool programme in urban Montreal, Rahm followed the trajectory of two teenage girls as they moved between school, home and community, over time. Rahm observed the girls' identity work in a science literacy programme and described how they carved out locations for participation as they actively positioned themselves through dialogue and writing. The girls sought out different pathways for participation: those avenues that were publicly valorised, as well as those that were marginalised. Rather than viewing 'multi-sited' ethnography as simply 'adding sites' of study to her research, Rahm argues for this approach to ethnography requiring new roles for the researcher as part of a 'collaborative imaginary' to view participants as they move through time-space. Rahm cautions educational researchers to bear in mind that as girls accrue sets of experiences, such experiences do not necessarily translate into increased agency or one's long-term participation in science but instead reflects the complexity of identity work.

Similarly, Tzou and Bell use critical spatial theory to examine the borders that demarcate 'place' among youth in an after school, environmental education programme in the western USA. The authors examined the discursive boundaries of fear and privilege in an environmental education curriculum surrounding what constitutes 'safe and unsafe' consumer beauty products and their impact on local water quality. As youth explored the range of meanings in this curriculum, Tzou and Bell describe one youth's 'counterscript' of his racial and ethnic heritage in after school programme, as he actively resisted the 'boundary-making' in the curriculum. This young man actively positioned himself on the margins of the activity as a location where he could express his 'Mexican-ness', while at the same participating in the curriculum. The authors argue that through ethnography educators can (re)design curricula that probe the dialectical relationship between the local production of scientific knowledge and larger socio-political narratives in environmental education.

To summarise, these articles illustrate the range of the pressing problems that characterise science education in rapidly changing global economy at a time when increasingly conservative ideologies are being 'mainstreamed' in public education. These articles express the many ways in which democratic participation in science education is being challenged. We believe the research assembled for this issue will be a resource for novice as well as more established scholars in science education, science-technology-society studies, and anthropology/sociology who wish to explore cultural studies of science in education. This collection of scholarship points to the complicated ways that larger political and governmental ideologies shape the ways youth and adults construct their understanding about the natural and physical world. Authors in this issue explored ways that ethnography opened up new understandings of the tensions surrounding science education reform using a variety of analytical frameworks. These articles point to promising directions for theorising the 'culture' of science education as we undertake educational reform. Indeed, ethnography is not only a valid approach for the study of science education; it is essential to the development of more equitable practices for instruction and learning.

Acknowledgements

The guest editors wish to thank Bob Jeffrey, the editorial board, and special issue reviewers for their comments during several stages of assembling this special issue.

Note

1. We would be remiss if we did not mention that several ethnographic monographs that have been central to an anthropological approach to science education: Eisenhart and Finkel's (1998) *Women's science: Learning and succeeding from the margins* and Nespor's (1994) *Knowledge in motion: Science, time, and curriculum, in undergraduate physics and*

management. Recently new ethnographic monographs have been published by Rahm (2010) and Long (2011).

References

Barab, S., and W.-M. Roth. 2006. Curriculum-based ecosystems: Supporting knowledge from an ecological perspective. *Educational Researcher* 35, no. 5: 3–13.

Barron, B. 2006. Interest and self-sustained learning as catalysts of development: A learning ecology perspective. *Human Development* 49: 193–224.

Berliner, D.C. 2002. Education research: The hardest science of all. *Educational Researcher* 31: 18–20.

Bronfenbrenner, U. 1979. *The ecology of human development: Experiment by nature and design*. Cambridge, MA: Harvard University Press.

Brown, A. 1992. Design experiments: Theoretical and methodological challenges in creating complex interventions in classroom settings. *The Journal of the Learning Sciences* 2, no. 2: 141–78.

Brown, A., and J.C. Campione. 1994. Guided discovery in a community of learners. In *Classroom lessons: Integrating cognitive theory and classroom practice*, ed. K. McGilly, 229–70. Cambridge: MIT Press.

Bruner, J. 1996. *The culture of education*. Cambridge: Harvard University Press.

Bybee, R.W., and B. Fuchs. 2006. Preparing the 21st century workforce: A new reform in science and technology education. *Journal of Research in Science Teaching* 43, no. 4: 349–52.

Calabrese Barton, A. 2001. Science education in urban settings: Seeking new ways of praxis through critical ethnography. *Journal of Research in Science Teaching* 38, no. 8: 899–917.

Carter, L. 2008. Sociocultural influences on science education: Innovation for contemporary times. *Science Education* 92: 165–81.

Cole, M. 1971. *The cultural context of learning and thinking: An exploration in experimental anthropology*. New York: Basic Books.

Collins, H.M. 1982. *Sociology of scientific knowledge*. Bath: The University Press.

Delamont, S. 1989. The fingernail on the blackboard? A sociological perspective on science education. *Studies in Science Education* 16, no. 1: 25–46.

Driver, R., and V. Oldham. 1986. A constructivist approach to curriculum development in science. *Studies in Science Education* 13: 105–22.

Eisenhart, M. 2006. Qualitative science in experimental time. *International Journal of Qualitative Studies in Education* 19, no. 6: 697–707.

Eisenhart, M.A., and E. Finkel. 1998. *Women's science: Learning and succeeding from the margins*. Chicago: University of Chicago Press.

Erickson, F. 1986. Culture difference and science education. *The Urban Review* 18, no. 2: 117–24.

Hammond, L., and C.B. Brandt. 2004. Science and cultural process: Defining an anthropological approach to science education. *Studies in Science Education* 40: 1–47.

Joslin, P., K.S. Stiles, J.S. Marshall, O.R. Anderson, J.J. Gallagher, J.B. Kahle, P. Fensham, et al. 2008. NARST: A lived history. *Cultural Studies of Science Education* 3: 157–207.

Kincheloe, J.L. 2001. Describing the bricolage: Conceptualizing a new rigor in qualitative research. *Qualitative Inquiry* 7, no. 6: 679–92.

Knorr-Cetina, K.D., and M. Mulkay, eds. 1983. *Science observed: Perspectives on the social study of science*. London: Sage.

Latour, B., and S. Woolgar. 1979/1986. *Laboratory life: The construction of scientific facts*. Princeton, NJ: Princeton University Press.

Lave, J. 1993. The practice of learning. In *Understanding practice: Perspectives on activity and context*, ed. S. Chaiktin and J. Lave, 3–32. Cambridge: Cambridge University Press.

Lave, J. 1997. The culture of acquisition and the practice of understanding. In *Situated cognition: Social, semiotic, and psychological perspectives*, ed. D. Kirshner and J.A. Whitson, 17–35. Mahwah, NJ: Lawrence Erlbaum Associates.

Lee, C. 2008. The centrality of culture to the scientific study of learning and development: How an ecological framework in education research facilitates civic responsibility. *Educational Researcher* 37, no. 5: 267–79.

Leont'ev, A.N. 1978. *Activity, consciousness and personality.* Englewood Cliffs, NJ: Prentice Hall.
Long, D. 2011. *Evolution and religion in American education: An ethnography.* New York: Springer.
Nespor, J. 1994. *Knowledge in motion: Science, time, and curriculum, in undergraduate physics and management.* New York: Routledge.
Osborne, J., and J. Dillon. 2008. *Science education in Europe: Critical reflections. A report to the Nuffield Foundation.* http://www.fisica.unina.it/traces/attachments/article/149/Nuffield-Foundation-Osborne-Dillon-Science-Education-in-Europe.pdf (accessed April 28, 2012).
Piaget, J. 1955. *The child's construction of reality.* London: Routledge and Kegan Paul.
Phillips, D.C. 2002. A guide for the perplexed: Scientific educational research, methodolartry, and the gold verses platinum standards. *Educational Research Review* 1, no. 1: 15–26.
Phillips, D. 2005. The contested nature of empirical educational research (and why philosophy of education offers little help). *Journal of Philosophy of Education* 39, no. 4: 1–21.
Rahm, J. 2010. *Science in the making at the margin: A multi-sited ethngraphy of learning and becoming in an afterschool program, a garden, and a math and science upward bound program.* Rotterdam: Sense Publishers.
Siry, C. 2009. Expanding the field of science education: A conversation with Ken Tobin. *Eurasian Journal of Mathematics, Science, Technology Education* 5, no. 3: 197–207.
Subramaniam, B., and M. Wyer. 1998. Assimilating the "culture of no culture" in science: Feminist interventions in (de)mentoring graduate women. *Feminist Teacher* 12, no. 1: 12–28.
Traweek, S. 1988. *Beamtimes and lifetimes: The world of high energy physicists.* Cambridge: Harvard University Press.
Tytler, R. 2007. *Re-imagining science education: Engaging students in science for Australia's future.* Australian Council for Education. http://research.acer.edu.au/aer/3 (accessed April 28, 2012).
Vygotsky, L.S. 1978. *Mind in society: The development of higher psychological processes.* Cambridge, MA: Harvard University Press.
Vygotsky, L.S. 1986. *Thought and language.* Trans. and ed. A. Kozulin. Cambridge, MA: MIT Press.
Weinstein, M. 1998. Playing the paramecium: Science education from the stance of the cultural studies of education. *Educational Policy* 12: 484–506.

Carol B. Brandt
Department of Curriculum, Instruction, and Technology Education
Temple University, Philadelphia, PA, USA

Heidi Carlone
Department of Curriculum and Instruction
University of North Carolina Greensboro, Greensboro, NC, USA

Unpacking 'culture' in cultural studies of science education: cultural difference versus cultural production

Heidi Carlone[a] and Angela Johnson[b]

[a]*Department of Teacher Education and Higher Education, School of Education, University of North Carolina at Greensboro, Greensboro, NC, USA;* [b]*Department of Educational Studies, St. Mary's College of Maryland, St. Mary's City, MD, USA*

In this article, we explore three anthropological approaches to science education research: funds of knowledge, third space/hybridity and practice theory. Definitions, historical origins, uses and constraints of each approach are included along with reviews of exemplary studies in each tradition. We show that funds of knowledge research draws on an earlier research tradition, cultural difference theory and rests on an assumption that groups build culture in response to fixed and static socio-political conditions. Practice theory is more flexible in that it allows researchers to study how groups create local meanings, which may conform to, resist or even transform those larger conditions through cultural production. We then illustrate the approaches by analysing the same case (that of a Mexican-American boy) using both cultural difference theory and practice theory, to illustrate the strengths and limitations of each approach.

Culture is, as Eisenhart (2001b) put it, an 'idea to think with' (16), one, which has been conceptualised, in a variety of ways by science educators. Historically, science education research has rested on psychological notions of learning and only in the past 25 years or so have anthropological perspectives surfaced with any traction. Primarily, these perspectives are borne out of a desire to reconceptualise science learning environments, to question taken-for-granted assumptions embedded in how the work of science education gets done and to bring to light processes of inequity and possibilities for challenging the status quo. One problem, however, is that science educators may not question or understand the nuances or historical roots of the concepts of culture they take up in their work. In this article, we explore three anthropologically grounded approaches to the study of science education, each with its own strengths and limitations: (1) funds of knowledge and its ideological forebear, cultural difference theory; (2) third space/hybridity; and (3) practice theory. We trace these approaches to their anthropological roots, unearth the understanding of culture in which each is grounded and explore what each approach reveals and constrains. To illustrate the strengths and weaknesses of various anthropological approaches, we end the article by analysing the school science experiences of a Mexican-American boy named Julio, using both cultural difference theory and practice theory.

Three anthropological approaches to science education

Funds of knowledge

In recent years, some science educators have used one conceptual model of culture, the shared knowledge and practices of a bounded group of people, to inform the 'funds of knowledge' approach, looking at ways to incorporate what students bring from their homes and communities into the science classroom. These efforts scaffold students' science learning by guiding them to realise that they already possess scientific knowledge and skills, and that their families and communities use science on a daily basis, whether or not they label it as such. The premise, according to Hammond (2001), is that 'linking school science to community activities is a way to enable all students to see science as accessible and relevant to their lives' (986).

The funds of knowledge approach was originally framed as a way of incorporating household knowledge and skills of students in Mexican communities in Arizona as a way of 'capitalizing on household and other community resources' to 'organize classroom instruction that far exceeds in quality the rote-like instruction these children commonly encounter in schools' (Moll et al. 1992, 132). In this study, anthropologists teamed with classroom teachers to conduct ethnographic research into the community and family funds of knowledge available to local students. As a result of this project, the authors 'learned that it is feasible and useful to have teachers visit households for research purposes', and that such visits could become 'the catalyst for forming research teams among the students to study topics of interest to them, or importance to the teachers, or for achieving curricular goals' (139). Funds of knowledge approaches were also grounded in concern for equity and often involved science. One researcher, herself an anthropologist involved in a project where Mien parents grew gardens and built a traditional garden house on school grounds, argued that 'science is so concrete that its potential as a hook is greater than that of most other discourses' (Hammond 2001, 997).

Basu and Calabrese Barton (2007) explained the funds of knowledge approach this way:

> Funds of knowledge refer to the historical and cultural knowledge of a community, . . . [and also] experiences and knowledge that may be more particular to a given family within the context of a community. . . . Funds of knowledge therefore include knowledge, action, and disposition or habitus with a recognition of how each of these domains are culturally constructed and refined. (468)

Or, more briefly, 'the cultural knowledge an individual possesses as well as how and why they act upon that knowledge' (485). Upadhyay described the benefits of a funds of knowledge approach for science learning, and its connection to equity:

> This greatly improves ethnic minority students' opportunities to learn and enjoy science and truly affords a social justice experience for these students by allowing immigrant students, like Hmongs, an opportunity to bring their cultural ideas into science classroom discourses. (2010, 58)

Researchers have been able to document positive outcomes associated with classrooms and programmes incorporating funds of knowledge approach. One teacher's efforts led to increased participation in school-based science activities by students and community (Upadhyay 2009). Students in an after-school science programme, which helped them draw on their funds of knowledge, were found to have 'pursued self-motivated science explorations outside the context of the classroom or used science in an ongoing way to improve, expand, or enhance an exploration or activity to which they were already deeply committed' (Basu and Calabrese Barton 2007, 469). Teachers in New Zealand who taught science classes with a high proportion of Māori students used Māori cultural knowledge of mountains during a landforms unit (Cowie, Jones, and Ortel-Cass, 2011). Cowie and colleagues found that both teachers and students benefited from the legitimate inclusion of Māori cultural knowledge in the science classroom; legitimising Māori students' funds of knowledge provided them an impetus to develop science-related identities in school science and provided teachers with deeper understanding of their students. Another programme helped Mexican-American parents of pre-schoolers identify science concepts and knowledge embedded in common household activities; the parents were in turn able to explain those concepts to their children (Riojas-Cortez et al. 2008). Science teachers in high-performing schools in Uganda were able to draw on students' everyday knowledge 'as a pedagogic resource or means to induct learners into science knowledge'; teachers in low-performing schools were more likely to focus on students' everyday knowledge *instead of* promoting science knowledge (Sikoyo and Jacklin 2009, 720). An ongoing professional development programme that taught teachers to incorporate students' funds of knowledge (among other techniques) yielded significantly higher student growth compared with similar schools whose teachers did not receive the training (Johnson and Fargo 2010). The funds of knowledge approach in this case consisted of the classic approach: 'conducting home visits to better understand how to bring real world experiences from different cultures into the classroom' (45). This last finding was particularly striking because schools were assigned to the treatment and comparison groups at random.

The funds of knowledge approach points to manageable, satisfying and effective actions that teachers can take to improve the science educations of their students. Further, it does so while framing students in non-deficit-based ways. Rather than blaming students for their failures in the science classroom, the funds of knowledge approach locates the failure in a mismatch between students and classroom, and points to concrete steps teachers can take to rectify that mismatch. It invites teachers to see students' households as 'containing ample cultural and cognitive resources, with great, potential utility for classroom instruction' (Moll et al. 1992, 134) and thus gives teachers concrete tools to 'use diversity as a resource in the service of learning' (Nieto 2000, 183). This approach 'contrasts sharply with prevailing and accepted perceptions of working-class families as somehow disorganized socially and deficient intellectually' (Moll et al. 1992, 134), the destructive and discouraging deficit view of cultural difference. As such, it is a dignified and respectful approach to teaching, and one with great power to improve science education.

Funds of knowledge and the cultural difference tradition

The funds of knowledge approach rests on a particular definition of culture: culture as 'patterns in a way of life characteristic of a bounded social group and passed down from one generation to the next', patterns which reflect a group's 'successful adaptation to relatively stable environmental (economic, social, and political) conditions' (Eisenhart 2001a, 210). Students' household funds of knowledge are derived from these adaptations. For instance, the students studied by Moll et al. (1992) had funds of knowledge that drew on the skills their parents used to earn money and care for children. In this, it draws on cultural difference theory, a body of educational research, which got its start in the 1950s. A fundamental assumption of cultural difference theory is that the culture of schools reflects the culture of the dominant class; what Delpit (1988) calls 'the culture of power'. Thus, for some students, the culture they experience at home is congruent with what they find in school; for others, however, success in school entails navigating a second culture consisting of less-familiar ways of organising time, space and language use. Within this logic, 'without special efforts to teach "culturally different" students the unfamiliar school culture, these students will, from the first day and through no fault of their own, have difficulty understanding what is expected of them in school' (Eisenhart 2001a, 210).

Cultural difference theory, as we argue above when describing the funds of knowledge approach, has been successful in framing the difficulties of poor children and non-White children in schools, without resorting to deficit-based explanations. It lets us understand some students' failures to thrive without blaming either the children or their teachers and schools; instead we can place the blame on miscommunication and cultural mismatches. This, in turn, has led to concrete changes in curriculum and instruction.

At the same time, there are limitations to understanding equity in science education with the cultural difference lens. First, it is grounded in the assumption that cultural forms are reactions to existing economic, political and social situations. Groups develop shared ways of using time, space and artefacts, and shared understandings in response to their conditions. The conditions to which these groups react are assumed to be fixed and static, affecting groups in unilateral ways. Such formulations assume enduring social structures without troubling their meaning and status. This fails to allow for the possibility that science itself could be changed and thus limits our thinking about transformative possibilities. The underlying theory of action of the cultural difference approach is to incorporate students' funds of knowledge into a science classroom as a bridge to science.

Viewed uncritically, one might assume that achieving cultural congruence in a science learning setting would leave no other obstacles to science equity. However, given what we know about the inequities in resource distribution among public schools in the USA, this is a cruel and inadequate assumption. It directs attention away from opportunity gaps in science; from the institutional processes which permit schools serving the most educationally needy students to be staffed with the least experienced teachers and given the fewest and oldest material resources (Education Trust, 2008, 2009; Kozol 2005; Lee and Orfield 2007; Shah 2011). It also fails to account for the processes by which certain kinds of people are anointed as science people, as belonging in science and with the right to use science, while other people are systematically produced as not belonging in science (Carlone and Johnson 2007; Johnson et al. 2011). And, it fails to account for the ways science as a discipline reproduces itself through culture, and the realities that a purely localised form of science (to which a culturally congruent science can fall prey) can be easily discredited because of its disconnect from larger networks of power (Carlone et al. 2009; Nespor 1994). In short, the funds of knowledge approach is vulnerable to what

Bourdieu calls the 'interactionist' error: 'reducing relations of power to relations of communication' (1999, 167). This formulation cannot account for how power gets reproduced or contested by long-term macro scale of political decisions about funding, meso-level institutional educational policies and shorter-term micro scale of myriad human interactions.

Third space

Researchers in the funds of knowledge tradition have responded to these limitations by studying the ways students use their funds of knowledge not just as a bridge to prototypical science but to bring new forms of science into existence. A third space or hybrid space is created when classroom members bring together elements of school culture and home culture to create something new. The concept of third space allows researchers to get at what people do with the funds of knowledge that students bring into the classroom; how teachers can make use of those funds and how teachers and students may be using them to spin-off from school science, making their own meanings. Third space work has focused on documenting that a third space exists between the 'official curriculum' and students' funds-of-knowledge-based 'counter-script' (Gutierrez, Baquedano-Lopez, and Tejeda 1999), and that this third space is pedagogically promising for educational equity, particularly in science. Calabrese Barton and Tan (2009), for instance, were able to show that when a teacher deliberately created opportunities where students were likely to construct third spaces, it led to greater student participation and better grades. Glasson et al. (2010), on the other hand, showed that science teachers in Malawi typically failed to take up rural elders' extensive knowledge of sustainability practices (including evidence-based reasons for using traditional rather than western growing practices) which could have served as a resource for a hybrid science, instead teaching science in a way which was 'largely irrelevant to most Malawian villagers' (Glasson et al. 2010, 138).

Third space/hybrid space researchers have also focused on developing the concept and related methods. Moje et al. (2004), for instance, enriched both the theoretical role of third space, describing it as 'a bridge, a navigational space, or a space for critical understandings of the relationship between science and students' "everyday worlds"' (54) and its aims to expand what is valued in science classrooms; to 'integrate, rather than divide, everyday and academic knowledges' and 'reshape how and why young people approach the content texts of their classrooms' (54). They also directed other researchers towards promising arenas to look for students' funds of knowledge: not just at family resources, but also as community and peer resources as well as popular culture resources. In doing this, they expanded the definition of culture beyond just 'patterns in a way of life characteristic of a bounded social group and passed down from one generation to the next' (Eisenhart 2001a, 210) to something which is no longer necessarily inter-generationally conveyed, nor shared only among identifiable groups.

Gutierrez (2008) further broadened the cultural assumptions underlying her use of third space, grounding it in social interaction. The creation of a collective third space, she argues, arises when 'students link the past and present to an imagined future and reorganize everyday concepts acquired through *social* interaction in *joint* activities into scientific or school-based concepts [our italics]' (158). In this formulation, activity in the third space is, in the words Ladson-Billings uses to describe culturally relevant pedagogy, 'specifically committed to collective, not merely individual, empowerment' (1995, 160). In third space, students and teachers can 'create opportunities to *collectively* generate new forms of joint activity to solve the double binds students encounter' (Gutierrez, 2008, 160). This use of third space is not only more overtly political than other uses, but it reflects the recent attempts among anthropologists to deal with the fact that 'conventional assumptions of culture as coherent and coterminous with social background, language use, region, religion, or ethnicity have become impossible to sustain' (Eisenhart 2001a, 213).

This realisation characterises an article by Calabrese Barton, Tan, and Rivet (2008), in which they examined the particular strategies a group of urban girls used to generate sanctioned or unsanctioned science-infused identities in science classes, without compromising their social identities. Although the article was grounded in hybridity, the authors did not make the girls' home funds of knowledge primary in their analysis. Instead, they directed their attention on the classic data of ethnographers: the things the girls said, did, made and used in science settings. The hybrid spaces in this article were no longer hybrids of students' family/ community funds of knowledge with the school's official curriculum; rather, they were places in which the girls could engage in what anthropologists call 'cultural production': 'a set of symbolic and material forms, affected but not determined by history and structure, actively appropriated or "produced" in groups to bring order and satisfaction to experiences' (Eisenhart 2001a, 212–13).

This approach has been criticised for framing hybridity as 'something teachers are to achieve' (Richardson Bruna 2009, 225). Richardson Bruna goes on to argue that educational researchers have often overlooked:

> 'Bhabha's critical insight that hybridity is always already there, regardless of whether the teacher decides he wants to 'create' it. Students' text-making activities and classroom experiences will always and necessarily be realised at the confluence of home and school literacy practices. (2009, 226)

The great use of hybrid space as a theoretical tool, according to this line of argument, is not to find out ways teachers can 'achieve' or 'create' it, but to identify:

> cracks in structure that may be taken advantage of by the subordinate group, in order to transform that structure. Learning the codes of the dominant group is a skill that can be put to use to dismantle the colonizing structure kept in place by that group. (Gonsalves, Seiler, and Salter 2011, 396)

In other words, the great use of third or hybrid space is to assist in studying power and identifying space for change: 'the ideas upon which any system of exploitation is built are necessarily contradictory and will produce gaps and spaces for resistance, transformation and liberation' (Shumar 2010, 502). We see this particular take on hybridity research as closely connected to practice theory, the third tradition in cultural studies of science education that we discuss below.

Practice theory

Practice theory is characterised by its attention to both micro- and macro-level factors, its focus on how individuals exhibit agency within the constraints of larger-scale structure and its (anthropological) focus on patterned, rather than individual, behaviour. Practice theory in educational research is rooted in ideas about cultural production (Eisenhart and Finkel 1998; Holland and Eisenhart 1990; Levinson, Foley, and Holland 1996; Willis 1977), situated learning (Lave and Wenger 1991) and networks of power (Nespor 1994, 1997). Users of practice theory often look at the way certain subject positions are celebrated or marginalised in a particular setting (Davidson 1996; Luykx 1996) and at the figured worlds within which those identities make sense (Holland et al. 1998). Thus, practice theory attends both to larger societal structures and to the ways individuals exhibit agency in everyday practices, working together to fashion cultural meanings that may reflect, contest and/or transform meanings implied by those structures. ‘Practice theory explores how individual and group cultures are formed in practice, within and against larger societal forces and structures’, and ‘gives equal weight to macro-level structures (i.e., gender, class, race) and micro-level factors (i.e., an individual's everyday practice)’ (Buxton 2001, 389). Within this approach, culture is not a set of practices handed down across generations as a response to enduring social structures, but a shared set of meanings, continually produced in everyday practice; new meanings can be congruent with what came before but also might have potential to alter what came before.

Practice theory attends closely to local processes, but the analysis cannot stay local. The theory directs user's attention to both micro- and macro-processes; it requires that researchers connect local practices and meanings to larger socio-historical meanings to which the local processes conform, resist and/or transform. Practice theory also attends to individual agency. Individuals can act to challenge prevailing cultural norms; however, they do so in the midst of pressure to ‘conform to the culture as currently practiced’ (Buxton 2001, 389). Within this framework, ‘individuals are not free to choose for themselves any view of the world, any way of acting in class, any definition of success, or any identity’ (Eisenhart 2001a, 215). The agency of individuals is constrained by structure, ‘the underlying principles that shape the normative patterns within social groups’ (Shanahan 2009, 45). When individual responses to prevailing cultural norms coalesce into ‘ongoing expressions of identity and purpose’ (Eisenhart 2001a, 217) shared by groups, they rise to the level of cultural productions.

Furthermore, to be indicative of cultural production, individual actions must reveal something about a group's shared meanings. Behaviour must be patterned, either in ways that align with local cultural meanings or contest them, to be of interest. Individual action must ‘be identified, not by individual statements of belief, but by patterns in the ways participants act in classrooms, label their own efforts, and describe themselves to others’ (Eisenhart 2001a, 217).

The strength of practice theory is its ability to resolve cultural puzzles, to explain surprising patterns and show how things are not as they seem. Margaret Eisenhart, whose work highlights the utility of practice theory in educational research:

> uses ethnographic techniques to describe both contexts within science education and science workplaces, and individual stories experienced by players within these worlds. She then illustrates how individual histories, attitudes, and identities of participants can enable them to rewrite what appear to be pre-determined stories. (Hammond and Brandt 2004, 17)

For example, in a study of a university science lab with a disproportionate number of women undergraduates, graduate students and postdoctoral researchers, Buxton (2001) was able to show that whereas one all-male group in the lab, the 'Cool People's Bay', had the strongest appearance of resisting the cultural norms of science (through their reputation for 'loud music' and 'wacky behaviour'), in fact that group was deeply engaged in conforming to wider cultural norms in science (particularly in their competition for resources). Another group, the 'Chick Bay', composed entirely of women, gave much less impression of resisting the cultural norms of science; however, they quietly undermined those norms much more than the researchers in the Cool People's Bay, by sharing scientific resources, training new lab members and tutoring struggling undergraduates:

> Whereas the behaviours in the Cool People's Bay exemplified a number of obvious and overt examples of resistance to the prototypical norms of science and scientist, there was much in the Cool People's Bay that conformed to those norms as well. In contrast, norm violations in the Chick Bay occurred in such subtle ways that they went largely unnoticed by the rest of the lab. Although much less noticeable than the boisterous behaviour of the Cool People's Bay, actions in the Chick Bay seemed more likely to lead to changes in science practice. (402)

In order to 'examine more deeply the call for a changed science curriculum', Carlone (2004) set out to investigate the experiences of girls in an Active Physics class. Active Physics is a thematic curriculum designed as a rigorous alternative to traditional, topic-centred physics curricula. Carlone 'wondered whether a different kind of school science, one that promoted alternative (and broadened) meanings of what it meant to "do science" and "be a science person", would make for a more inclusive and interesting science for girls' (393). What she found, however, was that even the girls who 'were successful and interested in the class (and who defined themselves as "lab" people) did not define themselves as "science people" and opted not to pursue further study in physics' (404). Carlone was able to make use of the simultaneous focus of practice theory on local meanings and larger socio-historical processes to show that this was happening because the setting was, despite its novel curriculum, characterised by prototypical meanings of science as 'difficult' and scientist as someone with 'raw ability'. She was also able to show that these local meanings, aligned with prototypical meanings, arose arbitrarily; the teacher found ways *not related to physics* to make the class more difficult (for instance, not giving students sufficient time to complete assignments).

In an example of the power of practice theory to understand the relationship between individual agency, societal structures, cultural production and cultural reproduction, Tonso (2006) showed how educational reforms at an engineering college failed to result in cultural change. Tonso engaged in participant observation between 1992 and 1996 at an engineering school that embraced a novel strategy for engineering education (requiring that students take not only traditional lecture-and-test engineering classes, but also design classes in which students worked in teams to solve real-world problems). She investigated the available, recognised engineering student identities, eventually collecting 126 different terms and focusing in on the 36 most common. In particular, she looked at the 'nerd' identity, especially the 'super-engineer nerd', students who 'excel at combining real-world practical knowledge with technical and scientific principles' (290) – the exact student who should shine in one of the design teams which were the central element of the school's reform strategy. She then went on to show that women students, even those who perfectly exemplified the characteristics of super-engineer nerds, were not recognised as promising engineers because of the gendered way that identities were understood at this school. She also showed that despite the school's commitment to producing hands-on engineers, students performing the nerd identity were not as valued as students performing another identity, the 'over-achiever', students 'whose test scores and grades ostensibly indicated being capable of doing engineering work at the highest levels' (302) regardless of how they actually performed in design teams. These two dynamics ensured that the school culture remained static; those students whose performances did not fit the higher-status 'over-achiever' identity became marginalised or left the community *even when they were competent engineers.* 'Ultimately, in spite of enormous student agency constructing subjectivities as engineers and performing themselves as engineers, and at times doing so in resistance to campus culture, little cultural change seemed likely to result from their creative performances' (301).

As these examples indicate, practice theory yields insights quite different from those of a funds of knowledge approach or the most common use of third space. Whereas funds of knowledge studies can provide teachers with guidance for how to draw on students' outside-the-classroom lives in order to make science meaningful and show them that they already have science knowledge and skills, practice theory facilitates the linking of in-class cultural norms to larger socio-historic patterns. Whereas third space studies tend to draw attention to particular moments in classrooms in which students and teachers blend official curriculum and student knowledge and interests to create a momentary hybrid science, practice theory focuses attention on *patterns* of production of new local meanings. Funds of knowledge approaches offer much more immediate guidance to teachers and science teacher educators; practice theory helps researchers and teachers understand how prototypical science meanings get endlessly recreated even in settings (like an Active Physics classroom or a reform-based engineering programme) where one might expect those meanings to be subverted, as well as insight into how science might gradually change (like a group of female graduate students establishing a norm of collaboration in their research bay). The kind of larger-scale insights offered by practice theory can be difficult to put into immediate practice; however, these insights are crucial for understanding how the whole system works, if one's interest is to dismantle that system.

Comparing cultural difference theory and practice theory: the case of Julio

To illustrate the strengths and limitations of cultural difference theory and practice theory lenses, we present a summarised case about Julio from Carlone's previous co-authored work (Rockford and Carlone 2011). Drawn from Carlone's larger research study (NSF #REC 0546078), Rockford and Carlone analysed this case using a cultural difference lens. Here, we summarise the case and that analysis, illustrating what understandings that lens enabled and constrained. Then, we re-examine Julio's case from a practice theory lens and its strengths and limitations. In doing so, we uncover taken-for-granted assumptions about culture that undergird both cultural difference and practice theory perspectives often overlooked in cultural studies of science education:

Julio is a shy student whose family emigrated from Mexico when he was in the first grade. He lived with his mom, his teenage sister and his sister's baby daughter. His family struggled financially, and he often took care of his baby niece after school. His fourth-grade teacher described him as having 'life resource skills' that others did not have: 'Drop him off ten miles from home, and he would safely find a route, get food along the way, make a friend, and find his way back home. The rest of the class would sit down and cry 'til somebody came and helped them'.

In 4th grade, Julio was playful, generally not afraid to take risks, not afraid to be wrong, observant, collaborative, and a problem solver. He struggled in other academic areas, but often shone in science, solving problems with electric circuits faster than others without calling attention to this fact and providing help to his friends by explaining concepts in Spanish. His teacher encouraged this practice, as the class was a dual-language immersion classroom for both Spanish and English. Often, he was the leader in science small group work, especially among his Latino male friends, encouraging them to stay on task, telling them what to write in their science journals and raising questions for them to examine in small groups. His excitement about science was contagious, especially among his Latino male friends.

However, his scientific performances could have easily been dismissed if you defined 'smart' in traditional ways. He did not make bids for recognition by expressing a large bank of scientific facts; neither was he the kid who engaged in robust whole-group discussions; nor was he the student whose journal was chock-full of carefully written observations and drawings. He was the anti-know-it-all student, the anti-competitive student. And yet, his fourth-grade teacher (Ms. Wolfe) and the classroom norms recognised and celebrated his understated and nonmainstream performances; Ms. Wolfe and some of his peers named him as one of the three smartest science students in end-of-year interviews.

Julio's sixth- and seventh-grade science classes (taught by the same teacher, Mr. Campbell) were dominated by worksheets, bookwork, and the occasional science project or cookbook laboratory activity (where students followed step-by-step directions to 'answer' a question). Here, you would most likely see Julio not talking, slouched back in his seat, his head down or propped up against his fist, often doing nothing, then copying classmates' answers. In these two years we observed Julio in science (sixth- and seventh-grades, roughly 20 observations), we never saw him volunteer any scientific answers, ask any questions, take any kind of leadership role in small groups, nor show any overt interest in any scientific topics presented in class. He often asked the teacher for the answers to worksheets, which often, the teacher provided. Over the two years of science, Julio grew more and more despondent, disinterested, and bored. His teacher said of him: 'If Jake or Edward are ambitious, Julio is the opposite of that. I think if you challenged him, he would say he's not very smart. It drives me nuts. I always kid him because he'll ask for the answers . . . All he wants is the answer. He just wants to check it off and put it away, because he doesn't think he is capable of even thinking'.

Analysis #1: cultural difference lens applied to Julio's case

A cultural difference lens directs one's attention to the relative match between the culture of Julio's home and the culture of the science class. The implication is that, when cultural norms, values and practices from home matched the cultural norms, values and practices from school science, he was more successful, interested and engaged. Rockford and Carlone (2011) analysed Julio's case by highlighting Latino cultural repertoires of practice (Gutierrez and Rogoff 2003) present in his fourth-grade classroom that were missing and/or misinterpreted in his sixth- and seventh-grade science class.

Consistent with a cultural difference lens, Rockford and Carlone (2011) argued that there are common, oppressive social, cultural, economic and political conditions Latinos in the USA face, and these oppressive conditions position them as:

> a racialised group subject to different types of racial discrimination. The backlash against Spanish and the 'English only movement' as well as the periodic ideological attacks—that are psychological, political, and economic in nature—on immigration provide clear evidence of this phenomenon that often manifests itself in violence. (Lynn and Parker 2006, 264)

There are specific Latino cultural repertoires of practice that can be harnessed as resources in such conditions. These are 'intergenerationally conveyed concepts, ways of talking, and belief systems that may be used and negotiated locally in communities that are often identified internally and by their neighbours in terms of ethnicity and race' (Gutierrez and Rogoff 2003, 21). For example, the Latino meaning of 'being educated' – or *educuación* – means something very different for Latinos than it does for Anglo families. In Latino culture, 'to be "educado" is to have a complex set of well-rounded understandings about the world and how to interact with others' (Rockford and Carlone 2011, 11) – somewhat akin to being well-mannered, someone who was 'raised right'. 'This is a very different than the Anglo version of "education" that usually equates with book knowledge' (Rockford and Carlone 2011, 11; drawing on work by Levinson 2001; Wortham, Murillo, and Hamann 2002). Another Latino family repertoire of practice is *familism*, where the comfort and good of the group is valued over that of the individual. This regard for the group and honouring of the kinship associated with it gets leveraged as a form of social capital (e.g., Villenas and Deyhle 1999). Closely related to familism is *humildad*, where 'children are taught to respect authority figures, and not single themselves out, in fact to actively work to take focus away from themselves in favour of the group, as a sign of respect and deference' (Rockford and Carlone 2011, 12 drawing on work by Ek, 2009; Valdes, 1996).

Rockford and Carlone (2011) analysed Julio's case by first highlighting the strong connections between the culture of his fourth-grade class and Latino values of educación, familism and humildad. For example:

> Julio positioned himself in ways that showed *educación*, by continually bringing others into science in ways that were deferential, held peers up as equals, and focused on group collaboration as the means to which he chose to 'do science'. Julio seemed to be unconcerned with the need to have the 'right answers', and instead positioned himself in ways where the members of his group could have experiences with scientific phenomena, almost as if taking a 'motherly' role in ways that resemble familism and kinship with his peers, especially during small group activities...These kinds of positionings are wrapped up in what it means to be 'bien educado' in Latino repertoires of practice relating to *educación*, as opposed to the dominant values of 'education', which are more knowledge- and fact-oriented. (22)

We demonstrated that the norms, values and practices in Ms Wolfe's (the teacher's) classroom recognised, drew on and celebrated practices that were consistent with Latino cultural values, thus enabling Julio's success:

> [W]hat was enabled for Julio in fourth grade, with the inclusion of Latino repertoires of practice, were opportunities to collaboratively interact, take risks, observe, and utilize

> *educación*, *familism*, *kinship*, and *humildad* in service of scientific inquiry that benefited Julio's positioning and the pursuit of scientific understandings that do not belong to a single individual, but to the group. (38)

Concurrently, Julio's heartbreaking decline in sixth- and seventh-grades can be similarly explained by a classroom culture that was not only inconsistent with, but stood in direct opposition to, Latino repertoires of practice. In these grades, competition, low-level tasks and individual bookwork dominated classroom routines and access to discrete scientific knowledge and 'perfect' (neat, 100% accurate) work held most sway. 'Ambitious' students were those who actively sought recognition for having the right answers, which violates the principle of *humildad*. The classroom culture of sixth- and seventh-grade science makes invisible Julio's abilities to be collaborative, to bring others into the excitement of scientific ideas and to help and get help from his Latino male friends as possible resources to enhance school science. In fact, the cultural mismatch between Julio's cultural values and school science's cultural values cast these potential strengths in a negative light.

Benefits of the cultural difference lens for Julio's case: what it enables us to see and understand

The cultural difference lens, when applied to the Julio case, provides a strong explanatory, non-deficit-based account of Julio's fourth-grade success and sixth- and seventh-grade difficulties in school science. The data about Julio's preferred ways of performing himself in school science fit perfectly with Latino cultural values described in the literature, such as *educación*, *famildad*, *kinship* and *humildad*. The cultural difference lens brings to the fore values and practices that may be traditionally overlooked in research studies and actively squelched in traditional school science settings. It allows us to 'see' Julio, and his successes and failures, differently than mainstream accounts of success and failure promote. For instance, because Julio was smart in ways that are often invisible with mainstream lenses, it would be easy to explain his difficulties in sixth- and seventh-grades by honing in on deficits with his academic preparation, reading level, writing abilities, motivation and/or home support systems. Instead, we now understand that Julio's repertoires of practice, when actively leveraged and celebrated as part of the class's cultural norms and values, become valuable assets for his own success and for creating a classroom community where knowledge is socially constructed, responsibility for learning is shared and authoritative knowledge is up for debate.

One can imagine practical applications from Julio's story as told through a cultural difference lens. A science classroom, like Ms Wolfe's, that recognises, leverages and celebrates Latino cultural repertoires of practice provides access for a wider range of student identities than those traditionally celebrated by school science. Teachers could read this account and immediately incorporate practices that make their classrooms more hospitable to students like Julio, practices that promote science as a collaborative, social endeavour and that push students to question authoritative accounts of knowledge. The account provides potential prescriptions for practice that also force us to question the ways traditional school science overlooks and silences non-mainstream students and shows us how we might capitalise on their cultural values and repertoires of practices to enrich and make more equitable school science.

Limitations of the cultural difference lens for Julio's case

Although the cultural difference lens offers potential, as is the case with any lens, there are limitations. Julio's account, told through the cultural difference lens, raises the possibility that, unless a class includes Latino cultural values like *educuación, famildad, kinship* and *humildad* and/or does not actively leverage cultural resources Julio brings to the setting, then Julio, and students like him who share similar cultural values, will not be successful. This seems rather deterministic and lacks nuance. It positions Julio as a passive recipient of the structures set forth before him. We are not ready to embrace this assumption, and we do not agree that the incorporation of Latino cultural values into classrooms is necessarily a prerequisite for or is the *only* way to achieve an equitable and inclusive science education for Latino students.

One reason why this assumption is troubling is that it ignores the possibility that Latino students, or students from other non-mainstream groups, might find science inherently interesting and an endeavour worth pursuing, whether or not it incorporates their cultural values. Our studies of women of colour who persisted in science (Carlone and Johnson 2007; Johnson et al. 2011) demonstrate that some women of colour in science author themselves in ways that leverage resources from their biographies, socio-political positionalities and moral commitments that allow them to navigate science educations and careers that did not always notice, let alone celebrate, their strengths and repertoires of practice. In other words, the cultural difference lens, at least as interpreted by the Julio case, limits the possibility of agency in transforming individuals' positions. On the other hand, we recognise that it was an injustice that the women had to author themselves in these creative ways that sometimes downplayed or denied aspects of themselves.

The cultural difference lens assumes that enduring features of history, like social class, racial, ethnic or gender stratification are primary or remain static over time. This is a benefit in many ways because it forces our attention to, and does not allow us to underestimate, the power of larger societal forces and structures in shaping everyday practices. At the same time, there is not as much room, with this lens, to explain or account for alternative, transformative or creative settings where race, class and gender are not primary determinants of culture.

Another limitation of the cultural difference lens for Julio's case is that it may assume that Julio's experiences would reflect all or even most Mexican immigrant student experiences in the same settings. Our research team followed seven other students from Ms Wolfe's fourth-grade cohort across 4 years of school science (two girls whose families were Mexican immigrants, one boy whose family emigrated from Ecuador, one mixed race boy whose father was from Colombia and whose mother was from Germany, one African-American girl and two white girls). While all students' science trajectories became increasingly difficult and troubling because of the problematic teaching practices and classroom culture in Mr Campbell's sixth- and seventh-grade classrooms (Carlone et al. 2011), none of them experienced the dramatic shifts in science engagement and learning that we witnessed with Julio. His case was extreme. Furthermore, the two other Mexican-American students and the Ecuadorian American student continued to earn good grades in science and did not regularly express the overt disaffiliation Julio did. In fact, of all the case study students, the one who was most consistently even with her identity performances in school science across all 4 years was a girl whose family emigrated from Mexico a few years before Julio's family did. In other words, she performed herself similarly in every context, whether or not Latino cultural repertoires of practice were included in the classroom culture.

A closely related problem with the cultural difference explanation of Julio's experiences is that the aspects of Ms Wolfe's fourth-grade classroom culture that enabled Julio to be recognised and celebrated as a 'smart science student' were not necessarily Latino-specific or Mexican-immigrant-specific. The practices that celebrated collaboration (which aligned with *famildad* and *kinship*) shared and socially constructed scientific knowledge (which aligned with *educuación*), encouraged students to question authoritative accounts of knowledge (*educuación*) and to show empathy and encouragement to others as they struggled to express themselves in a second language (*humildad*) worked for all students, not just the Latino students (see Carlone, Haun-Frank, and Webb 2011, for evidence of the classroom culture encouraging all students' science learning and affiliation).

We see here, also, the danger of essentialising Latino culture and/or with confusing culture with ethnicity. In the original manuscript, Rockford and Carlone (2011) acknowledged this risk and tried to guard against this interpretation of their work. To do so they drew on Ek (2009), who acknowledged the fine line between essentialising the identities of Latino/as and recognising them:

> as an ethnic group as bound together by a shared culture structure and a sense of a particular ethnic identity... It is important to point out that Latino/a students are agents who construct their identities within these larger macrostructural categories and labels... these categories... are meaningful to the lived experiences of Latinos/as in the United States in that these categories often affect their educational choices and opportunities. (407)

Even with these caveats, one could still critique the cultural difference lens for Julio's case as bordering on essentialism. A consideration of culture as *produced,* as is the case with practice theory, rather than *assumed a priori*, which is often the case with cultural difference lenses, minimises this risk more robustly.

Finally, issues of power are under-examined in a cultural difference lens for Julio's case. Power is acknowledged in as much as the oppressive social, political and economic circumstances that Latino/as in the USA face and in the ways schooling is not set up to recognise, leverage and celebrate Latino cultural repertoires of practice. In other words, power is constructed in a rather uni-directional, top-down way. The implied solution of Julio's situation, with a cultural difference lens, is the necessity of transforming school culture so that it is more aligned with Latino cultural values and practices. Doing so, we agree, would provide a powerful first step. However, we argue that step alone is not sufficient because of the many, multi-levelled issues of power inherent in society, schooling and science. To begin with, if we only pay attention to disrupting power locally, in a given classroom, we run the risk of advocating a school science that is *too locally situated* and disconnected from larger networks of power (Carlone et al. 2009). A too-local science, for example, might leave students feeling competent and affiliated in one setting only to discover that they are under-prepared for university-level science courses (Johnson 2007). Practice theory, we argue, allows us to look at power in more multi-faceted, multi-levelled ways.

Analysis #2: practice theory lens applied to Julio's case

Practice theory is adept at helping to explain why Julio became more alienated from school science than his classmates, drawing attention to the central puzzle of Julio. When we examine Julio's case through a practice theory lens, we pay attention to the ways local practices produce group-level meanings of 'science' and 'science person' that align with and/or contest larger socio-historical meanings. Further, this lens allows accounts 'for persons acting in cultural worlds' (Tonso 2006, 275). With this lens, we understand that what individuals learn and know depends on how social practices enable and constrain participation – in what times and spaces, with what tools, with what goals and endpoints (Margaret Eisenhart, personal communication, November, 1998). In this view, culture is affected, but not determined, by history and structure (Eisenhart 2001a). We ask: Who was he held accountable to be? How accessible, achievable and compelling were these implied identities and practices for Julio? How were those meanings similar to and different from what we see in traditional school science, which has a history of excluding and oppressing students from non-dominant groups? This lens toggles between structure and agency, micro- and macro-levels of analysis, and group-level and individual meanings.

Thus, to examine Julio's success in fourth-grade science, we examine the regularly occurring practices, and the accompanying implied meanings of science and science person, in the classroom. We cannot examine Julio's affiliation with fourth-grade science (i.e., his agency), without also examining the collective, patterned meanings of science and science person produced in the group (i.e., local structures). Furthermore, we understand the significance of Julio's affiliation better when we take into account historical, political, social and cultural meanings of *Latino male* participation in *US school science* (each italicised word carries with it its own powerful, enduring social, political and cultural history). In a slight shift from cultural difference perspectives, however, those socio-historical meanings are viewed as shaping and shaped by local practices and not necessarily as the enduring or primary influences on local practices and meanings.

So, in the fourth-grade (Ms. Wolfe's) classroom, scientific investigation was culturally produced as involving careful observation, question-asking and trying out one another's ideas together; it was a collaborative and generative endeavour (Carlone, Haun-Frank, and Webb 2011). Everyday practices of sharing scientific ideas with others; being patient and polite active listeners of one another's ideas; and being curious, critical thinkers implied a meaning of science that contested traditional socio-historical legacies of school science. These practices opened up spaces for participation and affiliation traditionally left closed for non-dominant students. The cultural meaning of 'smart science student' implied by these practices was made accessible to all students through consistent modelling, labelling and socially constructing expectations for what counts as 'good practice', and the teachers' unwavering expectations that all students perform themselves as smart (Carlone, Haun-Frank, and Webb 2011).

The purpose of fourth-grade science was not to arrive at pre-determined answers and fill students' heads with science content knowledge, but to produce students as careful observers of the natural world and natural phenomena, to think critically and to ask thoughtful questions that arose out of direct experiences with the natural world and natural phenomena. Doing so required that students embrace and perform the values of empathy, generosity, scepticism and collaboration. No one person could 'own' the knowledge; knowledge was shared.

The teacher did not solely create these cultural practices and values; one person does not create a culture. Students had to sanction, sustain, contribute to and participate in these practices in order for the class to achieve these cultural accomplishments. In Ms Wolfe's class, although students needed her direction, modelling, scaffolding and consistent reminders, all students aligned themselves (most of the time) with the cultural practices. Incidentally, these norms did not come 'naturally' to all students; many had to do significant identity work to align themselves with the norms, even though the norms were more inclusive than those implied by socio-historical legacies of school science. For example, Adam, another student in the class, often tried to invoke authoritative scientific texts to display his scientific knowledge. Ms Wolfe worked hard throughout the year to get him to question authoritative sources and to diminish the power of the 'know-it-all' student in her classroom. Adam may have been labelled as the 'smartest science student' in traditional school science contexts. But, in this context, many other students explained he was not as scientific because he was 'too technical'. This finding helps demonstrate students' buy-in and take-up of a broader meaning of science person.

The fact that Julio's teacher and some of his classmates recognised him, with his under-stated performances, as one of the smartest science students also demonstrates students' buy-in of a broader meaning of science student than would likely be the case with a classroom that reflected socio-historical legacies of school science. Again, the cultural accomplishment of contesting legacies of school science in this way cannot be over-stated; this was a significant accomplishment.

Julio's participation, in fourth-grade, indicated his approval and buy-in of the cultural practices and values and also demonstrated his roles in sustaining the classroom's cultural practices. He was persistent, collaborative, keenly observant, curious and had a love of discovery. In fact, his love of discovery was often infectious. As just one example, one day he made a discovery that, if the tank of African dwarf frogs was dark and left undisturbed, they were more likely to come to the surface to eat. After discovering this behaviour and communicating it excitedly to the teacher, the entire class came over to observe the frogs. His discovery and keen observations consistently played a role in sustaining the meaning of scientific investigation as collaborative, open-ended and dependent upon careful observation.

In sixth- and seventh-grade science, classroom work was dominated by worksheets, bookwork and the occasional cookbook lab and research project. Authoring himself as sole arbiter and authority of knowledge as well as primary 'entertainer', Mr Campbell positioned students as obedient recipients of his knowledge and/or as good audience members for his overtures to entertain. These practices implied very narrow meanings of 'good science student' and, in this case, the teacher had a lot of power in defining the classroom norms and values. To pull this off – because we know one person cannot create culture on his own – these norms and values drew heavily on socio-historical legacies of schooling. In other words, this way of 'doing school' made sense and was sustainable because it connected neatly with long legacies of teachers as authorities, students as passive recipients of knowledge and schooling as entertainment. Students had only limited agency regarding their participation. For example, they could perform: the obedient, diligent student; the admiring, appreciative student; the vocally disruptive student; or one who quietly opts out of participation. While many of his classmates chose the former two roles, Julio chose the latter role most often – quietly refusing to work diligently on his bookwork and sitting impassively during Mr Campbell's attempts at humour and entertainment.

To understand Julio's sixth- and seventh-grade performances from a practice theory lens, we examine the subject positions (meanings of 'science person') available to him and the ways he takes up, resists and/or transforms those subject positions. Why did other students, even those who experienced Ms Wolfe's excellent, student-centred pedagogy a couple years earlier, 'buy-in' to the limited subject positions and the narrowly defined classroom norms and values, while Julio did not?

One way to understand this is to draw on Willis' (1977) classic educational ethnography, *Learning to labour: How working class boys get working class jobs*, where he examined two groups of working-class boys in a British secondary school. One group – the 'lads', fully expecting to leave school at age 16 to pursue jobs requiring manual labour – openly and collectively resisted school authority, engaging in behaviours like fighting, drinking, smoking and 'havin' a laff' at teachers' expense. Unlike the 'ear-oles', the 'lads' did not buy into what Willis called the dominant educational exchange paradigm; students receive teachers' knowledge and school's sanctioned approvals (passing grades, diplomas) in exchange for acquiescing to teachers' control. They viewed the promise of school's rewards (a way out of their existing socio-economic status) as a false one and in not playing by the cultural rules of the schooling game they ended up in the same kinds of manual labour jobs their fathers occupied. Incidentally, the school's rewards were false; the 'ear'oles', who bought into the dominant educational exchange paradigm and believed in school's rewards, ended up in similar, unsatisfying working-class jobs.

Willis' work was so influential and speaks to Julio's case as viewed from a cultural production lens because his analysis considered the role of larger, societal and institutional structures in limiting, but not determining, agency. He was one of the first theorists to account for the production of culture, even though those cultural productions explained the lads' eventual social reproduction. Tonso (2006) explains:

> As such, the cultural level of day-to-day, go-to-school activities was constantly under production—where the lads were very active players, indeed—and the structural level of social class hierarchy penetrated into everyday social interactions (the cultural level) despite the school's progressive, liberal democratic ideals (at the institutional level). (276)

In Willis' account and Tonso's interpretation of it, the culture does not exist on its own; it is produced in practice, shaped by and shaping larger social and institutional structures. Further, we understand the lads' resistance as not only a *reaction to* practices they found suspect and/or unsatisfying (as might be the case with cultural difference theory) but, also, as the *production of* new meanings of schooling – of culture. At the same time, Willis' case of the lads demonstrates their agency in collectively creating new meanings of schooling that were more believable and rewarding (at least in the short term).

Perhaps Julio similarly rejected the dominant exchange paradigm of education to which classmates agreed. Although Julio did not express these sentiments exactly, we could initially infer that might have been the case with Julio's sixth- and seventh-grade science performances (and then follow up with additional data collection). The point is, with practice theory, we look upon Julio's different participation in fourth-, sixth-, and seventh-grade science in expansive ways; what does his resistance say about the struggle over the meaning of schooling and/or school science? Dominant meanings of schooling are not passively absorbed; they are 'fought over continually' (Giroux, 2006, 128). It is worth noting that Julio did not have similar opportunities as the lads to cause great ripples in local meanings of school science in Mr Campbell's classroom because he was not operating as part of a group who was resisting the institutional and classroom cultural practices and values. Willis argued, 'The basis for, and impetus of, this [cultural] production is the informal social group and its collective energies' (Willis 1977, 173).

Benefits of the practice theory lens for Julio's case: what it enables us to see and understand

A practice theory account, in summary, cannot examine Julio out of context from his socio-historical and political positioning, the socio-historical legacies of school science for Latino males, the cultural practices of his classroom and its implied meanings of 'science' and 'science person', and the ways he takes up, resists, transforms and/or negotiates these meanings, positionings and contexts. It is a multi-level, multi-faceted analysis that toggles between contexts and units of analysis. These are strengths of practice theory.

The practice theory lens assumes that culture is *produced* in practice. It does not assume, a priori, that students from certain ethnic, racial or geographic groups will act certain ways and embrace certain cultural practices or that their resistance, take-up and transformation of cultural meanings and practices will necessarily be distinct from those of their age-level peers from other groups. This addresses one shortcoming of the cultural difference lens; it minimises the danger of essentialising culture and treating all students from one ethnic or racial group as monolithic. Thus, with a practice theory lens, it is easier to examine within-group differences; we do not have to assume that all Mexican-American students responded in the same ways Julio did.

Practice theory provides an account of the production of culture *and* power. Everyday practices in Julio's sixth- and seventh-grade classroom produced a meaning of science and science person that were exclusive and narrow, shaped by and helping sustain larger, exclusive socio-historical legacies of school science and oppressive social structures. In examining these practices, it is easy to see school's ongoing disciplining of students (Foucault 1977) and the ways larger social structures get produced in everyday practice. Julio, through his unwillingness to play school science's games in sixth- and seventh-grades, gets framed as 'the opposite of ambitious' – i.e., as lazy. This meaning of Julio, as a lazy, Mexican-American boy reproduces, sustains and legitimises enduring social, historical, political and economic meanings. A practice theory lens examines the iterative production of power in everyday practice, rather than assumes that power operates in a top-down, uni-directional fashion.

Julio's sixth- and seventh-grade science stories, as told from a practice theory lens, account for limits to his agency and the ease of cultural reproduction. At the same time, the practice theory lens allows us to recognise transformative potential in Julio's story. For instance, everyday practices in Ms Wolfe's class produced Julio as smart and scientific. The fact that all students affiliated with some characteristics of the meaning of 'smart science person' (Carlone, Haun-Frank, and Webb 2011) implied that the class's practices produced a meaning of science accessible and believable for all students, quite a transformation from typical school science. This is a departure from the cultural difference lens, which can rest on an implicit assumption that enduring features of history are always primary or remain the same over time. Julio's success does not have to be explained based on his ethnicity or gender. Practice theory emphasises examination of how historical persons are formed in practice, within and against larger societal forces and structures which instantiate themselves in schools. We recognise Ms Wolfe's classroom culture as a radical accomplishment because we understand the incredible power socio-historical legacies of sorting and excluding inherent in schooling and science.

Limitations of the practice theory lens for Julio's case

The practice theory lens, as applied to Julio's case here, does not focus on what Julio brings to the setting in the ways that a cultural difference theory lens (in particular, in the ways a funds-of-knowledge lens) would. The lens focuses on his negotiations with and contributions to the culture of the setting as it gets produced, but it under-represents how the resources he brings to the setting affect the ways he will react to the setting. We note here, however, that this emphasis is not incommensurate with a practice theory lens. For example, Holland et al. (1998), using a practice theory lens, highlight the ways individuals leverage different resources in creative ways to enable new forms and meanings of participation in group settings.

Furthermore, the practice theory lens demands attention to multiple contexts and units of analysis – socio-historically enduring meanings (macro-structural), group-level meanings (cultural) and individuals' take-up, resistance and/or transformations of those meanings (individual). While we view this as a strength, it also is limiting in the time it takes to collect, analyse and interpret the data. Although ethnographic studies are time-consuming, a practice theory perspective often adds another layer or two of analysis in considering the iterative meanings produced by the individual and group, which shape and are shaped by macro-level structures.

The practice theory lens does not lend itself as well as the cultural difference lens to neat, prescriptions for practice. The emphasis is on understanding the meanings produced in groups, with the understanding that the outcomes (cultural accomplishments) are continually in question (Eisenhart and Finkel 1998), and can be imaginative and unpredictable. The 'local' nature of cultural production might limit the findings' generalisability to other settings.

Conclusion

In this article, we unpack implicit meanings of culture that pervade many cultural studies of science education. Table 1 summarises some of our key arguments about the strengths and limitations of different conceptions of culture for science education research. Current literature in science education (and educational research in general) is dominated by cultural difference perspectives, such as the funds of knowledge research we describe in this article, but also evident in the research about culturally relevant pedagogy. While the cultural difference lens provides strengths,

Table 1. Summary: cultural difference vs. practice theory.

Lens	Strengths: What it allows us to see	Limitations: What it overlooks
Cultural difference lens	Non-deficit-based explanation of Julio's success and difficulties Makes visible values and practices that undergird traditional science education and that may actively work against equitable science education Immediate applications to classroom practice that force acknowledgement and leveraging of students' strengths, backgrounds, values and life experiences	Implied solutions (i.e., to correct cultural mismatches) lack nuance Though it recognises the power of larger social structures, it assumes those structures remain static and/or are primary determinants of culture Assumes power is uni-directional Ignores possible intra-group differences and risks essentialising groups Implies different solutions for different cultural groups, making applications to a multicultural classroom difficult
Practice theory lens	Enables a multi-faceted, iterative analysis; from the micro- to macro-, from the group to the individual Assumes that culture is produced in everyday practice, allowing for a more nuanced perspective of intra-group differences and minimising risks of essentialising cultural groups Provides an overt examination of the ways power shapes and is shaped by everyday cultural practices Accounts for possibilities of transformation and helps illuminate limits to and possibilities for individual agency	Limits the focus on what histories, resources and practices students bring to new settings and newly formed cultural groups Time-consuming methodological endeavour to pay attention to multiple levels of analysis Lends itself to a more global perspective, which does not always enable neat prescriptions for classroom change

we argue that science education scholars have taken up this lens uncritically, without attention to its limitations. We challenge science education researchers interested in equity and who invoke culture as an analytic lens to lay bare the assumptions of culture underlying their work. The predominance of cultural difference lenses in science education research points to their uncritical adoption. Equity research needs to be mindful of blind spots; cultural studies in science education have not necessarily done so.

Acknowledgements

We acknowledge with gratitude the support from the National Science Foundation (Grant #REC0546078) to conduct this work. Any opinions, findings, and conclusions or recommendations expressed in this manuscript are those of the authors and do not necessarily reflect the views of the National Science Foundation. We also acknowledge Jean Rockford Aguilar-Valdez, who worked with Heidi Carlone on earlier analysis of the Julio case (Rockford and Carlone 2011).

References

Basu, S.J., and A. Calabrese Barton. 2007. Developing a sustained interest in science among urban minority youth. *Journal of Research in Science Teaching* 44, no. 3: 466–89.

Bourdieu, P. 1999. Language and symbolic power. In *Language and symbolic power*, ed. J. Thompson. Cambridge: Harvard University Press.

Buxton, C.A. 2001. Modeling science teaching on science practice? Painting a more accurate picture through an ethnographic lab study. *Journal of Research in Science Teaching* 38, no. 4: 387–407.

Calabrese Barton, A., and E. Tan. 2009. Funds of knowledge and discourses and hybrid space. *Journal of Research in Science Teaching* 46, no. 1: 50–73.

Calabrese Barton, A., E. Tan, and A. Rivet. 2008. Creating hybrid spaces for engaging school science among urban middle school girls. *American Educational Research Journal* 45, no. 1: 68–103.

Carlone, H. 2004. The cultural production of science in reform-based physics: Girls' access, participation, and resistance. *Journal of Research in Science Teaching* 41, no. 4: 392–414.

Carlone, H., M. Enfield, A. Johnson, J. Haun-Frank, and J. Kimmel. 2009. What you might expect and what you may have never considered to be culturally relevant science: Lessons learned from a multi-site ethnography. American Educational Research Association Annual Meeting, San Diego, CA.

Carlone, H., J. Haun-Frank, and A. Webb. 2011. Assessing equity beyond knowledge- and skills-based outcomes: A comparative ethnography of two fourth-grade reform-based science classrooms. *Journal of Research in Science Teaching* 48, no. 5: 459–85.

Carlone, H., and A. Johnson. 2007. Understanding the science experiences of successful women of color: Science identity as an analytic lens. *Journal of Research in Science Teaching* 44, no. 8: 1187–218.

Carlone, H., J. Kimmel, C. Lowder, J. Rockford, and C. Scott. 2011. Exploring the scope and limit of agency in the figured worlds of school science learning: A longitudinal study of students' identities. American Educational Research Association Annual Meeting, New Orleans.

Cowie, B., A. Jones, and K. Ortel-Cass. 2011. Re-engaging students in science: Issues of assessment, funds of knowledge and sites for learning. *International Journal of Science and Mathematics Education* 9, no. 2: 347–66.

Davidson, A. 1996. *Making and molding identities in schools: Student narratives on race, gender and academic engagement*. Albany, NY: State University of New York Press.

Delpit, L.D. 1988. The silenced dialogue: Power and pedagogy in educating other people's children. *Harvard Educational Review* 58, no. 3: 280–99.

Education Trust. 2008. *Core problems: Out-of-field teaching persists in key academic courses and high-poverty schools*. Washington, DC: The Education Trust.

Education Trust. 2009. *Education Watch National Report*. Washington, DC: The Education Trust.

Eisenhart, M. 2001a. Changing conceptions of culture and ethnographic methodology: Recent thematic shifts and their implications for research on teaching. In *Handbook of research on teaching*, 4th ed, ed. V. Richardson, 209–25. Washington, DC: American Educational Research Association.

Eisenhart, M. 2001b. Educational ethnography past, present, and future: Ideas to think with. *Educational Researcher* 30, no. 8: 16–27.

Eisenhart, M., and E. Finkel. 1998. *Women's science: Learning and succeeding from the margins*. Chicago, IL: The University of Chicago Press.

Ek, L.D. 2009. It's different lives': A Guatemalan American adolescent's construction of ethnic and gender identities across educational contexts. *Anthropology & Education Quarterly* 40, no. 4: 405–20.

Foucault, M. 1977. *Discipline and punish: The birth of the prison*. Trans. by A. Sheridan. New York: Vintage Books.

Giroux, H.A. 2006. Disposable youth, racism, politics of zero tolerance. In *America on the edge: Henry Giroux on politics, culture, and education*, 175–88. New York: Palgrave Macmillan.

Glasson, G.E., N. Mhango, A. Phiri, and M. Lanier. 2010. Sustainability science education in Africa: Negotiating indigenous ways of living with nature in the third space. *International Journal of Science Education* 32, no. 1: 125–41.
Gonsalves, A.J., G. Seiler, and D.E. Salter. 2011. Rethinking resources and hybridity. *Cultural Studies of Science Education* 6, no. 2: 389–99.
Gutierrez, K.D. 2008. Developing a sociocritical literacy in the third space. *Reading Research Quarterly* 43, no. 2: 148–64.
Gutierrez, K.D., P. Baquedano-Lopez, and C. Tejeda. 1999. Rethinking diversity: Hybridity and hybrid language practices in the third space. *Mind, Culture & Activity* 6, no. 4: 286–303.
Gutierrez, K.D., and B. Rogoff. 2003. Cultural ways of learning: Individual traits or repertoires of practice. *Educational Researcher* 32, no. 5: 19–25.
Hammond, L. 2001. Notes from California: An anthropological approach to urban science education for language minority families. *Journal of Research in Science Teaching* 9, no. 38: 983–99.
Hammond, L., and C.B. Brandt. 2004. Science and cultural process: Defining an anthropological approach to science education. *Studies in Science Education* 40, no. 1: 1–47.
Holland, D., and M. Eisenhart. 1990. *Educated in romance.* Chicago, IL: University of Chicago Press.
Holland, D., W. Lachiotte, D. Skinner, and C. Cain. 1998. *Identity and agency in cultural worlds.* Cambridge: Harvard University Press.
Johnson, A. 2007. Unintended consequences: How science professors discourage women of color. *Science Education* 91: 805–21.
Johnson, A., J. Brown, H. Carlone, and A. Cuevas. 2011. Authoring identity amidst the treacherous terrain of science: A multiracial feminist examination of the journeys of three women of color in science. *Journal of Research in Science Teaching* 48, no. 4: 339–66.
Johnson, C.C., and J.D. Fargo. 2010. Urban school reform enabled by transformative professional development: Impact on teacher change and student learning of science. *Urban Education* 45, no. 1: 4–29.
Kozol, J. 2005. *The shame of the nation.* New York: Crown Publishing.
Ladson-Billings, G. 1995. But that's just good teaching!. *Theory Into Practice* 34, no. 3: 159–65.
Lave, J., and E. Wenger. 1991. *Situated learning: Legitimate peripheral participation.* Cambridge: Cambridge University Press.
Lee, C., and G. Orfield. 2007. Historic reversals, accelerating resegregation, and the need for new integration strategies. Civil Rights Project [*Proyecto Derechos Civiles*], University of California, Los Angeles.
Levinson, B.A. 2001. *We are all equal: Student culture and identity at a Mexican secondary school, 1988–1998.* Durham, NC: Duke University Press.
Levinson, B.A., D.E. Foley, and D.C. Holland. 1996. *The cultural production of the educated person: Critical ethnographies of schooling and local practice.* Albany, NY: State University of New York Press.
Luykx, A. 1996. From Indios to Profesionales: Stereotypes and student resistance in Bolivian teacher training. In *The cultural production of the educated person: Critical ethnographies of schooling and local practice*, ed. B. Levinson, D. Foley, and D. Holland, 239–72. Albany, NY: State University of New York Press.
Lynn, M., and L. Parker. 2006. Critical race studies in education: Examining a decade of research on US schools. *The Urban Review* 38, no. 4: 257–90.
Moje, E.B., K. McIntosh Ciechanowski, K. Kramer, L. Ellis, R. Carrillo, and T. Collazo. 2004. Working toward third space in content area literacy: An examination of everyday funds of knowledge and discourse. *Reading Research Quarterly* 39, no. 1: 38–70.
Moll, L.C., C. Amanti, D. Neff, and N. Gonzalez. 1992. Funds of knowledge for teaching: Using a qualitative approach to connect homes and classrooms. *Theory Into Practice* 31, no. 1: 132–41.
Nespor, J. 1994. *Knowledge in motion: Space, time and curriculum in undergraduate physics and management.* Washington, DC: Falmer Press.

Nespor, J. 1997. *Tangled up in school: Politics, space, bodies, and signs in the educational process.* Mahwah, NJ: Lawrence Erlbaum.

Nieto, S. 2000. Placing equity front and center: Some thoughts on transforming teacher education for a new century. *Journal of Teacher Education* 51, no. 3: 180–7.

Richardson Bruna, K. 2009. Jesús and María in the jungle: An essay on possibility and constraint in the third-shift third space. *Cultural Studies of Science Education* 4, no. 1: 221–37.

Riojas-Cortez, M., M.E. Huerta, B. Bustos Flores, B. Perez, and E. Riojas Clark. 2008. Using cultural tools to develop scientific literacy of young Mexican American preschoolers. *Early Childhood Development and Care* 178, no. 5: 527–36.

Rockford, J., and H. Carlone. 2011. Silencio en ciencia: A longitudinal case study of Julio's silencing in school science. Annual meeting of the National Association for Research in Science Teaching, Orlando, FL.

Shah, N. 2011. Federal data shed light on education disparities. *Education Week* 30, no. 36: 29.

Shanahan, M.-C. 2009. Identity in science learning: Exploring the attention given to agency and structure in studies of identity. *Studies in Science Education* 45, no. 1: 43–64.

Shumar, W. 2010. Homi Bhabha. *Cultural Studies of Science Education* 5, no. 2: 495–506.

Sikoyo, L.N., and H. Jacklin. 2009. Exploring the boundary between school science and everyday knowledge in primary school pedagogic practices. *British Journal of Sociology of Education* 30, no. 6: 713–26.

Tonso, K.L. 2006. Student engineers and engineer identity: Campus engineer identities as figured world. *Cultural Studies of Science Education* 1, no. 2: 273–307.

Upadhyay, B. 2009. Teaching science empowerment in an urban classroom: A case study of a Hmong teacher. *Equity & Excellence in Education* 42, no. 2: 217–32.

Upadhyay, B. 2010. Middle school science teachers' perceptions of social justice: A study of two female teachers. *Equity & Excellence in Education* 43, no. 1: 56–71.

Valdes, G. 1996. *Con respeto: Bridging the distances between culturally diverse families and schools.* New York: Teachers College Press.

Villenas, S., and D. Deyhle. 1999. Critical race theory and ethnographies challenging the stereotypes: Latino families, schooling, resilience and resistance. *Curriculum Inquiry* 29, no. 4: 413–45.

Willis, P.E. 1977. *Learning to labor: How working class kids get working class jobs.* New York: Columbia University Press.

Wortham, S.E.F., E.G. Murillo, and E.T. Hamann. 2002. *Education in the new Latino diaspora: Policy and the politics of identity* Vol. 2. New York: Ablex Publishing Corporation.

From science as 'content' to science as 'interpretive key': experiences and reflections from a science course in teacher education

Laura Colucci-Gray and Christine Fraser

School of Education, University of Aberdeen, Aberdeen, UK

Despite science's central role in European culture, public perception of, and participation with, science is characterised by contradictions and conflicting agenda. School curriculum reform, for example by Scottish Government, promotes 'science for citizenship', yet teachers' understandings of the nature of science and its relationship with society are often underdeveloped. This article reports on the experiences of a group of first-year students enrolled in the course 'socio-cultural perspectives in science learning' within a primary teacher education programme. Drawing on sociocultural theory, attention was given to language as a system for constructing symbolic realities. Through participation in activities, which explored science knowledge production and communication, students were encouraged to reflect on the nature of science and the dynamic relationship between content, methods and value-frameworks. Findings cluster around interconnected themes: students' views of knowledge, language genres across communities and knowledge–power negotiations. The course attempted to move from a space for knowledge consumption to an intermediary space between science and society for students to inhabit in the dual roles of citizens and co-constructors of knowledge. In this transition, linguistic genres pointed to the complexity of interactions between curricular materials offered by the course and newly established social networks. The article argues for further consideration of linguistic practices in science education as a means for disclosing interpretations and cultural framings and a necessary step in equipping teachers to understand 'science for citizenship' issues.

Introduction

In recent decades, we have witnessed a paradoxical situation. On the one hand, the increasing rate of consumption of natural resources and concurrent degradation of the environment has endangered the survival of millions of people as well as numerous non-human species. On the other hand, we are confronted with the dominant view of unlimited economic growth in which scientific and technological knowledge plays a crucial role.

Recent work from the field of Science and Technology Studies has underlined the need to develop awareness of the role of science and technological knowledge in shaping people's identities as consumers of knowledge and nature in the neo-liberal market. Technologies are conceptualised as 'products of social relationships that are not politically innocent' (Takeshita 2010, 55). As western modern science has come to constitute itself as a privileged form of knowledge production to support global economic development, the inequities experienced by large groups of people in terms of meeting basic, fundamental needs raise questions about the modus operandi of a knowledge enterprise that has induced profound transformations of natural environments and social relationships, with unpredictable and unforeseen consequences. More than ever before, a concern for equality on a finite Planet goes hand in hand with a model of knowledge production that is open to participation and pluralism (Jasanoff 2007), in a way that we propose is both ethically and epistemologically necessary.

Such a turn in science studies challenges content-centred discourses in science education to look at science as a social practice: views of science are translated into political choices, social relationships, international development strategies (Söderbaum and Brown 2010) and indeed educational approaches (Colucci-Gray et al. 2006). We argue that a context of science learning that values participation and personal involvement is necessary to bring forth a culture of science that is open to other forms of knowledge and that is 'interpretive' rather than descriptive, that is, enabling the expression of a diversity of experiences, languages and points of view. The aim of this article is to research the social and communicative dynamics of such a context and to uncover the constraints in promoting such a shift.

Background

Over the years, we have been teaching on undergraduate courses in science education aimed at prospective primary teachers. Chris was a former primary teacher in England while I had entered University teaching through a research path, after being in Italy first and subsequently having moved from England to Scotland. While different in terms of experience and age, as women who entered academic life we both found ourselves in a space of critical interrogation of the values and practices of knowledge production in western society and particularly the academy's embrace of an exclusionary model of science, closed to the different, that being the feminine or just 'the other', outside the sociocultural boundaries of academia (Harding 2008).

The majority of our students enter the University straight from school; they are mainly young girls, of 17–18 years of age intending to pursue teaching as a professional career. For these students, the encounter with science is almost inevitably a fearful moment. Science appears as a foreign, alien subject. It is a common occurrence, for example, when we touch upon the concept of 'electricity' at any given workshop to find students feeling uncertain and in discomfort as they try to give meaning to the word used, often by looking for a material representation of an abstract notion. Yet, 'electricity' is something which we experience every day, from its origin and uses to its effects in the environment and society. Electricity indeed has influenced greatly the construction of women's identity over time continuing to shape the lives of our student teachers as women in western societies (Takeshita 2010).

The repeated experience of alienation from science, however, is far from being only a reality for our students, who are white, young females. Rather, a wealth of studies has reported the separation from school science experienced by students across the board of language and ethnicity. As reported by Krugly-Smolka (1996) science as it is presented in school 'is not for everyone' and this exclusionary feature of science stretches beyond the acquisition of what may be difficult, abstract or unfamiliar content. Rather, people who engage routinely with the content of science are those who are able to enter science as a particular domain with its own basic categories, artefacts, symbols and beliefs – indeed a 'culture'. Hence, the tools and language of science respond to a specific way of positioning oneself within a social group and in relation to the natural world; from this perspective, however, for many students science remains something 'other' which can enter into conflict with other background cultures, school and family culture, or ethnic cultures. As such, while science affects and shapes life opportunities in our societies, the construction of scientific knowledge inhabits a knowledge space that is removed from the reach of students' everyday concerns and foci for action.

Aims of the study

This study resulted from our desire to understand the context of learning about science that we were able to provide to the students and the desire to move from a space for knowledge consumption to an intermediary space between science and society, for students to inhabit in the dual role of citizens and co-constructors of knowledge.

Here, we introduce the concept of approaching science 'in an interpretive key' to indicate a process of transition from the learning of science as the acquisition of a body of knowledge to learning science as part of a process of cultural interrogation of its practices and worldviews, indeed 'an interpretation'. This process equates to enabling students to give meaning to the way in which science has depicted the natural world and to connect those descriptions with other sets of meaningful interpretations, stemming from the articulation of values and experiences.

The question explored through this experience would thus focus on: What images of science are held by students approaching the course? How far does the linguistic interaction within the course enable students to develop an identity of knowledge constructors?

Before beginning this analysis, we give an introduction to the policy context followed by a review of current developments in the epistemology of science supporting a theme of citizens' empowerment. Drawing on Latour's notion of continuity between localising and globalising activities in the production and maintenance of the social (Latour 2005), we argue that the course was a space of interpersonal and linguistic practices that connected the class to wider institutional and social relations which are imbued with power. Understanding such practices and our efforts as tutors to engender students' participation was thus important to recognise the struggles for the legitimisation of students as knowledge producers that come with the need to produce a more empowering form of science.

The policy context: the public understanding of science and science for citizenship in the UK

According to the Expert Group Report (2008, 9) '[...] the general public fears science because of its risks'. In response to this state of affairs, within Europe, science education is receiving greater attention. In January 2008, a report prepared for the Nuffield foundation in England states:

> Science is an important component of our European cultural heritage. It provides the most important explanations we have of the material world. Teachers who are knowledgeable and effective communicators are able to engage their students in substantive conversation; ask searching questions; and have a deep understanding of their own subject. (Osborne and Dillon 2008, 9)

Such a position recognises the epistemic power of scientific inquiry and the legitimacy and acceptability of its explanations. Teachers are invested with the responsibility of making science more approachable and intelligible for students and so to contribute to the Public Understanding of Science (Millar 2002) and more generally to the promotion of responsible citizenship (British Council 2010). A similar view appears also in the recently launched 'A Curriculum for Excellence' in Scotland, which is promoting the development of broader civic capacities, active learning contexts and enhanced cognitive capabilities (Education Scotland 2012). Within this new curriculum, science learning is central to enabling citizens to access Science and Technology and contribute to the country's economic growth.

While this commitment to citizenship by the government is overwhelming in scale and it should certainly be welcome, a number of authors have also argued that the discourse of the Public Understanding of Science with its emphasis on the 'active' citizen has been driven by a deficit model of the public who is supposedly having problems with 'understanding science' and therefore it displays negative attitudes towards it (Nisbet and Scheufele 2009). Participation and contribution from the public in this sense is construed as gaining the public's support for science.

If this is the idea of science that dominates and it is perpetrated at all levels through the educational and social systems, citizens, public administrators and politicians are encouraged to put their trust in science in order to make private and public decisions: the knowledge of science is considered to be certain, because it is based upon principles of experimental testing in controlled conditions and objective, because it is free of value-choices and subjected to standardised procedures of quality assurance. Once validated, this kind of certain and objective knowledge is expected to filter directly into the normative sphere to inform the correct course of action, a position that is still widely encouraged in decision-making processes (Stirling 2010).

The knowledge society and the transformations of science

A caveat to this position is that science credibility has gone though important changes in recent times. The accelerated development of the four new techno-sciences (nano-, bio, info- and neuro-cognitive) reflects the increasing power of science to isolate the fundamental components of the natural world and to enable the reconfiguration of living systems to suit human needs and priorities, at a local and global scale. At the same time, however, research and implementation are not confined to the controlled and simplified conditions of the laboratory but they are carried out directly on the socio-environmental systems of the planet. Citizens' lifestyles and values are located alongside a set of risks which are unknown and difficult to manage. Such changes in the nature and practices of science have stimulated a lively debate in the epistemology of science. The preferred approach of conventional science is analytical and quantitative, aimed at describing the complexity of the living world through the fundamental laws of chemistry and physics. This approach is accompanied by disciplinary specialisation and a growing difficulty in communicating results and findings not only to the public but also between experts from different disciplinary affiliations. According to Longino (2002), science is now characterised by a plurality of theoretical explanations of the living world, each one confined within well-defined spatial and time boundaries.

Furthermore, Longino (2002) recognises controversy as a reflection of the plurality of levels, points of view and framing schemata that characterise scientific inquiry and which followed almost as a necessary response to the inherent complexity of the natural systems under study. Groups of researchers may come to produce experimental results which are scientifically correct but which may be in contradiction with each other. In such conditions, it becomes apparent that learning about science cannot be simply defined as 'understanding science better' or increasing the cognitive capability for assessing the credibility of scientists' claims.

The understanding of complexity has indeed brought together a number of suggestions for re-thinking the idea of science and the position of humanity in the world. A complex structure makes it impossible to identify specific boundaries between processes and phenomena even when operating at different scales in space and time: the micro is connected to the macro and present conditions of a system are rooted into systemic relationships developed over a long period of co-evolution (Volk 2003). Hence, modern techno-science generates some sort of 'paradox': as we proceed with our analysis, we expand both the level of our knowledge as well as the range of our ignorance (Kelly 2008). As reported by Harding (2008), in current issues of science, technology and society reliability is not enough. She argues, 'Reliability of conventional reliable knowledge has been criticised as producing far too narrow understandings of nature and social relations' (83). Hence, knowledge is incomplete not simply because it can be replaced by other interpretations and new knowledge but because such knowledge trespasses traditional disciplinary boundaries and escapes the control of its peers: 'the epistemological core of science is actually crowded with a variety of norms and practices' (84). This point leads us to an understanding of language and social practices in the production and communication of scientific knowledge.

The importance of language

Language is a system of signifiers (words, symbols and images) operating as interconnected elements to create meanings or *signifieds*. So all that we build that is new in our knowledge and understanding is grounded in the system of signs that we have at our disposal (Camino and Dodman 2009).

Also in science, new experiments and measures, new methodologies and new contexts almost routinely contribute to change and/or expand existing conceptions and language changes accordingly. So for example, the linguistic evolution of the word 'gene' (Fox Keller 2000) points to the process of objectifying portions of the genetic material by means of a particular signifier, forming the epistemological foundations of biotechnological sciences, its distinction from cognate fields and the beginning of a particular scientific discourse for describing the living world. In turn, as reported by Harding (1994), cultures become repositories for historically developed knowledge and beliefs about nature and western science are a cultural system amongst many, developed and interpreted by means of language.

Paradigmatic changes are also made possible through the production of new concepts and metaphors which are expressed by means of language in association with the existing culture. As reported by Ravetz (2003):

> Surprisingly, we find metaphor embedded in the most apparently scientific sorts of discourse. Darwin's theory of evolution relied explicitly on metaphors, as the 'Tree of Life'; and what is 'Natural Selection' but the greatest metaphor of them all? (Ravetz 2003, 68)

By operating from a language perspective, it becomes easier to overcome the image of a dogmatic science which describes the world as it is. Rather, there is recognition that knowledge production occurs in a performative setting whereby knowledge and knowing cannot be separated from questions about the ethical, cultural and normative features of knowledge production contexts (Wieser and Karner 2010). So, with the interrogation of scientific language – from the roots of its words to its pictorial and diagrammatic representations of natural phenomena, we gain access to broader cultural representations of time, space and our own views of the world in a humanistic way (Matthews 2000). From this perspective, learning science in an interpretive key will involve the development of a linguistic awareness recognising that language is both affecting and being shaped by the nature of our thinking and interaction with other people and with the natural world (Camino and Dodman 2009).

So how do we prepare teachers for approaching science in an interpretive key?

Many studies in science education literature have focused on providing evidence of teachers' and students' epistemological beliefs (e.g., Abd-el-Khalick and Ledermann 2000; Apostolou and Koulaidis 2010). Drawing largely on questionnaires surveys, findings point to students' inadequate views of the nature of science and teachers' lack of confidence towards the discussion of uncertain and controversial science (Gray and Bryce 2006). Amongst the approaches being proposed, in the UK, Matthews (2000) argued for the contextualisation of scientific topics within the history and philosophy of science, while other strategies included the creation of learning environments sharing the characteristics of authentic research environments, where students can practise the skills of academic debate and argumentation (Osborne 2010). Conversely, Quale (2002) reminds us of the role of the hidden curriculum and the implications of naive views on people's perceptions about what constitutes relevant and legitimate knowledge in institutionalised environments. For example, Lee (2002) and Brandt (2008) focused on situations where guidance was given to students in recognising continuities and discontinuities between academic, western science and local, everyday experiences in one's own culture. From this perspective, Aikenhead (1996) advocated for teachers taking on the role of culture-brokers, facilitating students' travels across borders into the sub-cultures of science. According to Aikenhead (2004), this way of learning science corresponds to a process of 'anthropological' learning or personal sense making. Academic bridges are put in place to allow for a degree of personal exploration, including academic abstraction, analysis and self-initiated participation. So, participation within a social space requires linguistic awareness and the ability to recognise the values, norms and beliefs associated with the way language is used. If conventional studies in science education have depicted the deficiencies of students' preparation, forms of linguistic ethnography have provided insights into the development of students' voice in the production of knowledge on matters concerning science (e.g., Furman and Barton 2006). However, there is a paucity of studies looking at the development of beginning teachers' identity in academic contexts and how the social and linguistic dynamics between teachers and students can empower students to negotiate border crossings amongst a multiplicity of voices and languages.

The context of the study and the course

This study focuses on the networks of communicative interactions evolving as part of a course seeking to introduce a group of beginning teachers enrolled in the Bachelor of Education Programme at a UK University to a view of science in dynamic interaction with cultural beliefs, norms and languages and open to other forms of knowledge.

The course was titled 'Socio-cultural perspectives on science learning' co-taught by Chris and Laura – co-authors of this article. The course had a duration of 12 weeks with an allocation of 40 hours contact time and a nominal 40 hours for preparation and assessment. Concurrent with University curricular reform, the Bachelor of Education programme changed the original provision of courses offered by education staff to encourage first- and second-year students to engage with a diverse range of subjects within the University. This course was thus newly created and offered as an elective option open to other disciplines. As it transpired for that year (and indeed in the following years) participants were all Initial Teacher Education (ITE) students.

Besides, while the teaching of history and philosophy of science have featured in UK science education research, they do not feature prominently in ITE in Scotland. Science courses for ITE students adopt a social constructivist approach, dominant in science education for the last 30 years (Jenkins 2000), and introduce the idea of alternative conceptions in science, yet without sustained investigation of the nature of science.

Our course design drew on sociocultural historical activity theory (Engeström 1987) in recognising that participants (students and tutors) with diverse prior experiences use language as a tool to achieve a shared understanding in a joint purpose. In doing so, a community, in which work is shared and rules made, is formed. Developing such an approach, however, is difficult because of the need to take full account of the learners' experiences with a tradition of science which prioritises content knowledge, a common pattern of secondary school teaching across Europe (Apostolou and Koulaidis 2010).

Furthermore, the course was located within a nested system with the institutional sphere on the outside and the newly created community of learners – with their expectations and social norms – on the inside. This acquired history of rules, task allocations and sense of community that accompanied each participant into the course would result in participants having diverse experiences of language and communication in 'scientific' contexts. This recognition required a shift in perspective from learning science as a static body of knowledge to learning science as a dynamic, participatory and contextualised process. Hence the notion of science as an interpretive key became the keystone of the course and informing strategies for the co-generation of knowledge through collaborative communicative processes, akin to Levin and Greenwood's (2001) notion of action-research process in which 'the diversity of experiences and capacities is an opportunity for the enrichment of the research/action process' (105). Students were made aware of their role of participants in the action-research process and the data-generating nature of the activities offered. Appendix 1 provides a sequence of the main activities covered in the course. Particularly in this article, we focus on three, selected examples of learning science in an interpretive key: the visit to the zoology museum, 'plant classification' and 'webs and chains'. All three activities were linked to the idea of classification of living forms based on selected criteria. Every classification, however, is a theory about the creatures it seeks to describe (Gould 1998) and students were encouraged to familiarise themselves with a diversity of classification models and their pedagogical messages: a critique of the singular criterion models for example allowed for representations that included features of the whole organism. While this choice of criteria defeated the purpose of a manageable classification, the representation so constructed sought to promote awareness of the complex, diverse and interconnected nature of life on the Planet. An extract from Amartya Sen's 'Identity and Violence' was then adopted as a means to bring the epistemological reflection into other disciplinary realms and further interpretation of the cultural dimension of science.

How students became involved in the process of interpretation and the nature of the social context underpinning their participation was thus a concern of the research.

Methods

As indicated by Hammersley (2006) this experience departed in many ways from traditional ethnographic studies, for the nature of the community we were studying was by necessity a temporary one, centred upon a particular activity – the course. This condition implied the existence of clearly defined boundaries – both in time and space – that would prevent the observation of long-term processes or patterns concerning students' learning and identity formation to direct attention instead to the specificities of communication dynamics and rules of that community.

Nevertheless, the course enabled observation of meaning-making practices generated by course activities designed to involve students in epistemological reflection while taking account of the broader social, cultural and historical context of students' experiences with science learning. This, in turn, provided insights into people's epistemological categories and value assumptions while promoting a more contemporary view of science within the traditional academic contexts.

To this purpose, the study involved a form of linguistic ethnography as described by Rampton (2007) recognising that (a) 'meaning takes shapes within specific social relations, interactional histories and institutional regimes, produced and construed by agents with expectations and repertoires that have to be grasped ethnographically' and that (b) meaning is more than just expression of ideas, as it is extensively signalled in the linguistic and textual fine-grain (585). Indeed as indicated by Kats and Csordas (2003), the inclusion of phenomenological constructs of mood, climate and non-verbal communication is important in capturing the nature of meaning-making processes.

Data were gathered through students' free responses at the start and end of the course (see Appendix 2 for the questions for the final evaluation) and through their mid-point reflective essays. Concurrently, tutors' own reflections and observations were collated systematically at the end of every lesson to provide a commentary on the purpose of the lessons and how this was interpreted within the social dynamics of the class. The analysis was undertaken by general qualitative inductive methods (Hood 2007), and looking at the intersection of students' images of science and their involvement in the course. The paths of three students – Leah, Rya and Solange – are provided as examples of alternative learning experiences and the mediating role of language in supporting students' interpretations and the sharing of learning purposes.

Findings

Views of knowledge and language boundaries across communities

> [Notes from Laura's reflective journal]. It is the first day of the course. As a relatively new lecturer in the School I am keen on getting to practice with building a course that aims high, that tackles dominant scientific discourses and prepares young teachers to integrate a critical analysis of science in their profession. With such feelings and motivations I introduced myself to the students and present the course as aiming to explore and discuss ideas of science in society, in a collective fashion. I stressed the need for working together and to establish personal relationships in order to venture into a common experience.

At the start of the course, the pattern of relationships amongst the participants was laid out. I was presented with a group of 25 students most of whom do not yet know one another. The group included also two Spanish students – Rya and Solange – who are currently in the UK on the Erasmus exchange programme. From a linguistic point of view, I could be perceived as a 'foreigner' by native students, my speech being coloured by a Southern European accent which positions me closer to the two Spanish students. My surname, however, includes an English component and in the eyes of the students I am clearly located within the University as a knowledge broker within the Institution. From a social status point of view, a distinction clearly existed between the tutor on the one side and the students on the other. We start by introducing one another using an identity card activity followed by a social game to encourage people to find out a more about one another and establish a common experience (Engeström 1987). Students appeared hesitant at the beginning, but by the end of the activity they seemed to have relaxed, and somewhat expectant towards knowing what the course is set out to do. I draw on the ice-breaking activity to introduce the idea of community and the rules of participation. In this spirit, I ask students to share their expectations on the course. I have a clear objective as to what I want this course to be – a forum for reflection and inquiry. Perhaps unsurprisingly, however, students' understandings of the rules and purposes of the course are quite different. A core theme in their writing is the expectation of coming to the course to *take* something away: 'What I can take from the course which I can use in my career in teaching' and 'What resources would be better and where to find them'. Students' expectations for their own learning were set out on gaining knowledge to transmit to others gaining skills for knowledge transmission: '[...] confidence and skills in presentations so I can talk confidently to a class' and 'To learn how to present to a group: to be a teacher involves presenting to a class in a professional way'.

Students' language points to the existence of commodified ideas of knowledge in which a clear division of labour exists between knowledge 'givers' and knowledge 'receivers' mirroring conventional ideas of science 'communication' (Nisbet and Scheufele 2009).

My declarations about science as an opportunity for learning through personal contributions and participation were contrasted by students' previous experiences of science learning and the pressures to perform under the assessment system 'I liked biology but I did not do well in exams', an aspect that they referred to as well in relation to this course: 'A lot of effort and participation is required on my part and my elective is no exception. I discovered that my work is continuously assessed and that group work, presentations and participation are more assessed elements of the course'. Hence, the purpose of the course, envisaged by tutors, was challenged by students' expectations of the course as being a means to a measurable end ('To enjoy myself and to pass'). This difference in purpose, which reflects social interests and economic pressures within a student culture, encourages strategic development, and conversely neglect, views of knowledge akin to Harding's (2008, 58) 'patterns of systematic knowledge and ignorance'.

However, student culture is not homogeneous. For some students, the beginning lecture stimulated thinking about the nature of science: 'I would like to learn the theories that concern science and society which were mentioned in the introductory lecture as I always thought science was not that type of subject' and the role of science in society: 'To learn how science impacts communities'.

Students are also aware that linguistic and discursive abilities are necessary to satisfy the requirements of the course and negotiate access within academia: 'Being able to contribute to tutorials and discussion as this is required throughout the programme', and in the social environment of the class: 'To discuss and put my views and opinions across to the rest of the class'. For Rya and Solange the appropriation of discursive abilities equated to the acquisition of a basic, human skill that would enable them to operate in a cross-cultural encounter: 'To improve my English as this is not my mother tongue'.

A theme of empowerment and disempowerment began to emerge regarding students' abilities and expectations about themselves as learners and knowledge producers within the course. In this respect, I was conscious of my use of the English language: terms such as nature of science; body of knowledge; learning agenda, were introduced according to the codes and practices of a particular *speech* community (Kramsh 1998) that may still be alien to the students.

Communicative obstacles would inevitably shape students' responses to the course activities. Wilson (1992) provided a close examination of the relationship between a fieldworker people in the community in international settings: 'At any rate, people in all sorts of problematic and asymmetric situations clearly do strike up genuine friendships of enormous richness. But such relationships can still remain pregnant with tensions' (192). Hence, the patterns of relationship amongst participants and the *language genres* characterising such patterns were key elements to the purpose of learning science in an interpretive key.

Students' views of science

Following the introductory lecture, Chris and I were concerned with how to stimulate students to think about the relationships between science and society and their own personal position as 'consumers' of scientific knowledge and nature (Takeshita 2010). Referring to the work of Cole and Knowles (2010), the purpose of pooling information from participants in the form of personal narratives and memory maps was a way to record the history of the learning process shared by the community of students and tutors. 'The Draw a Scientist' (DAST; Mason, Kahle, and Gardener 1991) and the Views on Science-Technology-Society instrument (Aikenhead and Ryan 1992) were used as a stimulus for generating students' images and perceptions.

Students' drawings of a scientist were analysed and discussed together starting from the scores on a rating rubric adapted from Mason, Kahle, and Gardener (1991) looking for observable characteristics (Table 1). Overall, students subscribed to the stereotypical model of scientists which has persisted since the 1950s and namely that

Table 1. Frequency of characteristics in students' DASTs.

Characteristics	Frequency ($N = 18$)
Personal characteristics	
Male	18
White/Caucasian	18
Laboratory coat	17
Eyeglasses	15
Pens/pencils in pocket	9
Bald or unkempt hair	18
Other (blemishes, facial hair, lipstick, etc.)	2
Practical tools	
Test tubes	18
Flasks	18
Bunsen burners	8
Solutions in glassware	18
Machines	1
Patient on bed	1
Knowledge tools	
Books	2
Filing cabinets	8
Friends and colleagues	2
Computers	0
Research journals	0
Other (signs charts, periodic table, etc.)	4
Environment	
Laboratory	18
Office	1
School/university	2
Outdoors	0
Overall appearance	
Eccentric	17
Sinister	3
Positive	14

of a white male wearing eyeglasses and a laboratory coat replete with pens in the top pocket (Figure 1).

While there have been suggestions that students' images are not necessarily stable over time and may be sensitive to environmental elements (Losh, Wilke, and Pop 2007), the students' drawings pointed to a certain ambiguity about scientists who are imbued with the power of accessing Nature's core elements but who also seem to deviate from

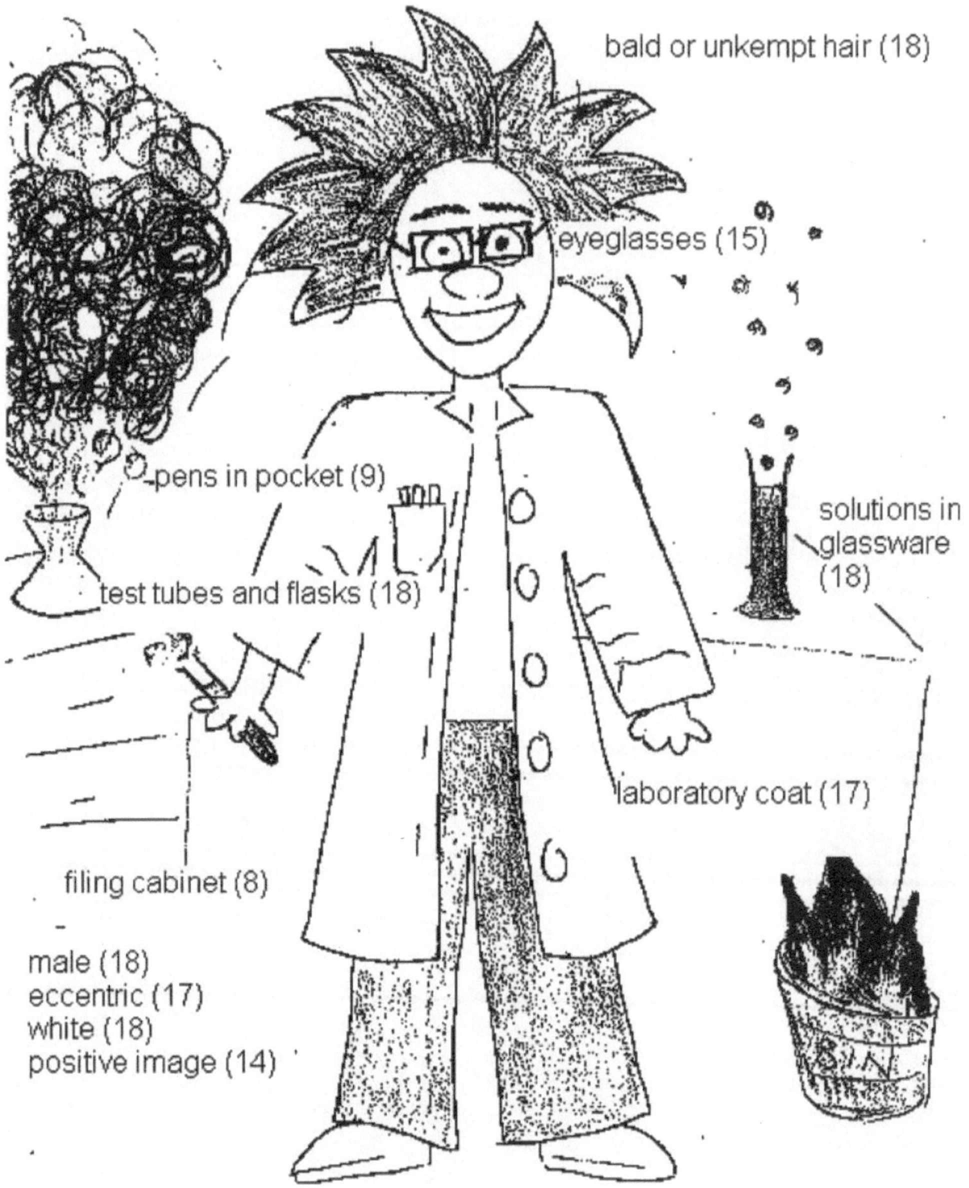

Figure 1. Typical student response to DAST (numbers in brackets indicate frequency across the sample).

'normality'. The exaggerated and caricature-like features of their drawings conveyed a sense of cultural, social and political separation between the worlds of the scientists and the worlds of the students in terms of symbols and behavioural practices.

The discussion of selected items from the VOSTS questionnaire was thus an opportunity to begin to ask questions about the works of science. Two items (60,311 and 91,011) related to the influence of religious views on a scientist's work and the role of 'invention' in the production of scientific facts. On the first item, students espoused the view of a 'neutral' science clearly separated from the world of religious belief although some students acknowledged the potential influence of religion on the 'discoveries' that a scientist makes. However, when interrogated on the role of invention in the production of scientific knowledge, students fully converged on the idea that evidence 'speak for itself' and the application of the correct methods will lead to a singular description. 'So where do the problems we experience with scientific-related developments come from'? – I dare to ask. I can see Rya and Solange starting an animated discussion with each other. The 's' sound of their accent makes people smile and turn their heads in their direction. When I ask everybody to write down some open questions they would like to address in the course of the 10 weeks, they address the critical interaction between science and policy: 'Who should take the responsibility, the scientists or the policy makers'? Whereas other groups focused more narrowly on the cognitive and knowledge generation process of science: 'Does the constant discovery of new ideas some time [sic] have a bad effect? Can you discover too much'?

The activity seemed to have engaged students' in debating the nature and purposes of science. An idea of science that produces truths appeared in contrast with the value-ladenness of knowledge and actions in everyday life (Harding 2008) and the possibility to extend the constituency of informed opinion outside government and scientists. The entry in Laura's reflective journal for that day conveys certain 'pushiness' towards trying to shift students' ideas:

> Today's activities reveal much of what is a very traditional conception of science. I cannot help feeling unnerved about these ideas being so widespread. I feel the urge to push students [...]; give them a role of inquirers into the places for science communication, stimulate their thinking about science as a 'presence' in our culture, which informs our language, ideas about nature and actions.

Interpreting science?

In week 6 into the course, students were invited to inquire into the process of science communication by visiting a museum, learning centre or an art gallery of their own choice, with the specific task of observing the interplay of words and images in the communication of science. As it turned out, all students chose to visit the zoology museum which was conveniently located on campus. Students made their visit in groups with a view to communicate their experiences to their peers as a part of a class presentation. So far our efforts to use the activities as mediums for sharing knowledge within the newly formed community had sat somehow in between a form of researcher-driven agenda and participatory inquiry as we were preoccupied with gathering students' drawings and writings by having to remind students to share their work. The visit to the museum was the first opportunity for students to take an active role in putting knowledge and experiences across to the rest of the group.

In addition, a writing task was set up for students to report on their experiences of science communication and begin to engage with the process of writing.

In line with University's regulation for continuous assessment, the writing would be marked and would count for 20% of the overall mark, the remaining 10% being peer-assessment of group presentations and 70% for the final essay. Perhaps unsurprisingly, the idea of assessment was problematic for the students. Some students appeared in distress; one of the students, Leah, became tearful. The week following their presentations, a sour mood penetrated the course. Students were surprised by the task and expected very clear instructions about the content of the essay.

Chris and I were struggling with finding a way to deal with the students' needs: 'I was amazed at the lack of understanding of the task. I was even more surprised by the "what has zoology got to do with science" comments'. [Tutors' reflections]. We had set out the visit to the museum as a form of exploratory learning, yet the assessment seemed to have 'changed the purpose from finding out about the museum to presenting the information to "pas". Needing to pass raises issues of insecurity [. . .]?' [Tutors' notes].

We experienced an overwhelming urge to play safe and retreat into our traditional roles of knowledge guardians and knowledge transmitters. We were also conscious that the episode had made an impact on the relational climate within the course and Leah appeared to be particularly difficult to engage.

In effort to understand the nature of the contradiction, we involved students in a participatory exercise to elicit their feelings as 'offers' written on post-it notes of different colours and laid out on paper plates. Students were encouraged to respond to stimuli on their emotional disposition towards the visit to the museum, for example 'I was surprised by [. . .]'; 'I was disappointed in some way [. . .]'; 'a question I want to ask [. . .]'. It emerged that the majority of the students had never been to a zoology museum before; as guests or 'tourists' in a new place (Aikenhead 1996) they were expecting to be guided towards knowledge. They admired the large-scale displays of the Bengal tiger and the Tyrannosaurus Rex but they struggled to extract meaningful knowledge: 'I was disappointed about the lack of information on the exhibits. Felt I could have learnt more'; 'By the lack of people (none) around to ask about the museum display and a little by my lack of ability to link it with what I was doing in class'; 'Amount of information – lack of, too much'. In contrast to the majority of the class who struggled with understanding what was to be learnt from the visit, for Rya and Solange the visit to the museum enabled them to make observations about the analytical features of scientific thinking: 'I was able to see how the different types of birds were classified'; 'I learned about the way in which preserved animals can allow us to have a real view on how these animals look'; 'I was also surprised by the similarity in humans and apes' (evolution).

For students who are familiar with the culture of school but unfamiliar with the culture of science, and somehow akin to the 'smart kids' described by Aikenhead (1996), the presentation of science was just an overwhelming display of artefacts, while the epistemological practices of zoology as a scientific discipline remained obscure and in contrast with their own values; for example they were puzzled by the display of natural items in 'cases' and the purpose of observing 'stuffed animals'. Their use of language did not carry awareness of the ontological distinction between considering something as a common object (a 'stuffed animal') or a 'scientific specimen' functioning within a theory or empirical framework (Abd-El-Khalick and

Lederman 2000). For students embarking on ITE studies, the view of a unitary science had implications for their perceptions of what legitimately counts as 'science', as exemplified by the following quotation from a reflective essay: 'However, after reflecting back with the group I discovered that I had learnt something [. . .] Zoology was a science, this is something that I had not been aware of because of my narrow minded view of science, being chemistry, physics and biology'. [JR-para. 19]

From knowing that to knowing why

At this point during the course Chris and I were questioning how we could support students with the interpretation of science to facilitate the crossing of boundaries between the realm of science and that of everyday life and help students redraw the images of themselves as knowledge producers. Following up from the visit to the zoology museum, the three weeks before the end of term were dedicated to the theme of taxonomy and its continuities with the general culture (Ravetz 2003). We began from the observation of plants and the cognitive and practical processes involved in configuring a life object as a sample or 'specimen' for botanical classification. In this particular case, we were concerned with reflecting on the duality of any classification system which identifies unique features by cutting across a web of relationships. Subsequently, we introduced students to a reflection on different classification systems and how they have been used to produce 'narratives' about the organisation of the natural world. The idea of a world of separate kingdoms appeared very familiar and appealing to the students, whereas the more recent model of metabolically shared paths and synergistic interactions (e.g., Volk 2003) was more difficult to grasp. Students were drawn to the observation of morphological characteristics while they lacked familiarity with the physiology, ecology, genetics of living organisms and how the multi-dimensionality of our knowledge could allow for more fluid and inclusive views of the natural world. The last activity before the Easter vacation was dedicated to drawing students in the process of sharing basic knowledge of living organisms and how they are commonly classified to build a web of relationships that had relevance at both the cognitive and symbolic levels. The activity presented organisational and cognitive challenges; and it was the behaviour of one student in this task, Leah, which illuminated the different ways in which students shared in the purpose of the course and the fragile network of relationships that had been formed (Engeström 1987).

Inclusion of exclusion?

After a brief introduction to food webs and the description of the activity, each student was given two cards showing named organisms, their task being that of establishing prey–predator relationships and then, using a ball of coloured yarn, to connect the organisms to build up the food web. Leah felt in 'holiday mood', stated that she did not want to be in class, and had to be cajoled into taking part. Throughout the activity she kept on giggling, dropping the ball of yarn and disrupting the representation of the food web.

Halfway through the activity, the tension was palpable: the activity was progressing slowly because of students' weak subject knowledge and Leah's repeated interruptions. I was tempted to ask Leah to leave, but Chris moved her into the role of Sun. She continued to giggle and to make comments yet she kept hold of the threads for the remainder of the activity. In the performance of self Leah was engaging with (Goffman 1956) a declaration was made about how she perceived the rules of the context to be: when approaching the holiday period, the contract of give and take between tutors and students is expected to get looser. However, if in the 'school' context her expectation was that other students would join in the disruption, in the course, there was little or no uptake of Leah's communicative act of disruption: most students ignored the behaviour or signalled their annoyance. At the end of the workshop Leah, accompanied by her friend Mia, complained to the tutors about having been intimidated by the Spanish students' speaking in Spanish and their unfriendly glances, but when the tutors pointed out that her own inappropriate behaviour had possibly antagonised other students, Leah began to cry.

This episode further reinforces the gap between the norms the course sought to create and students' sociocultural experiences of school learning (Engeström 1987). More generally, Leah's reactions towards the Spanish students points to the difficulty of reshaping interaction genres within an institutionalised learning space in a way that supports a shift from transmission to personal and collective interpretations (Quale 2002).

Closing remarks

Towards the beginning of this article we posed two questions. Discussion of the responses to DAST and VOSTS instruments earlier in this study suggests that students embarking on this course, in common with others, held stereotypical views of scientists (Figure 1) and viewed science as neutral, value-free and separate from other disciplines. While disappointing, given the recent arguments for teaching science in context (e.g., Matthews 2000), this was not surprising. Of greater interest is the impact of the course.

Collating students' perceptions of their experiences at the end of the course (see Appendix 2 for the full list of questions) show that they found it 'confusing' (nine indications) but also 'interesting' (six indications) (Figure 2). These descriptions reflect the uncertainty and stimulation associated with crossing cultural borders between everyday life and science (Aikenhead 1996). Clearly some students managed this transition better than others and indicated changes in perspectives: 'I am able to relate this activity to society's views and how they can change over time according to

Figure 2. Adjectives generated for 'How did I live the experience of this course?'

the science research available'. However, students also expressed disappointment about the divergence from their initial expectations, particularly in relation to the lack of teaching strategies to implement in class, and the predominance of theory versus practical work. This suggests that students and tutors did not share a common purpose (Engeström 1987), and point to an underlying theme of separation between theory and practice.

Successful negotiation of borders requires skilful interpretation of language, customs and context. However, from the examples provided, we suggest that student participation in the linguistic genres was both enabled and constrained by course conditions.

Some students recognised that the collaborative, exploratory style was conducive to self-expression, discussion and personal reflection: 'Group participation especially the pesticides activity, the creative side of it'; 'The tutors were very easy going so it was easy to participate and discuss' and 'There were never right or wrong answers'. However, students' diverse prior experiences resulted in differential access to the social networks and student–student interactions within the class, as evidenced by the food web activity. Similarly, expectations that their role was to 'receive' rather than 'provide' knowledge highlighted inequalities within the classroom genre: 'The lack of participation from rest of class hindered my learning in class. Not really sure why. In some ways if no one spoke it made it difficult for me to speak out loud'. Seemingly, the rules of participation were either not understood or not enacted (Engeström 1987).

In turn, tutors' abilities to adapt to students concerns, while holding fast to underpinning principles of inclusion and participation, further impacted on student and institutional interactions. While the freedom to create this course was enabling, the institutional framework within which it operated was constraining. For example, teaching and assessment practices created tensions which prevented students from accessing the performative settings of the investigations, group presentations and mid-point essays.

So those two words – interesting and confusing – are emblematic of a tension that continued to exist within the students' learning experiences in relation to the opportunities for border crossing (Aikenhead 1996).

Those students with prior experiences of crossing cultural borders (Rya and Solange) travelled furthest on the journey towards independent knowledge construction whereas those with more restricted experiences (Leah) left the course with some way to go. Yet all students are expected to become primary schoolteachers, so one might ask whether the ability to become independent constructors of science knowledge really matters to students. Clearly, classification and interdependence are key areas of knowledge about the natural world. However and arguably more importantly, they can also be viewed as metaphors for diversity and complexity within society (Ravetz 2003). It is the understanding of the social, cultural and historical heritage of society which underpins our notion of citizenship.

To prepare teachers for science for citizenship would equate to contributing to the creation of a culture which does not silence and alienate but proactively seeks a multiplicity of voices. However, such an approach would need to include a more explicit interaction with the natural world beyond rational forms of academic language (Osborne and Dillon 2008) to espouse 'other styles of communication (e.g., visual methods, storytelling) which allow people to express their particularities and collective identities in a more embodied way' (Söderbaum and Brown 2010, 189).

This may help reframing the idea of practice beyond what were expected to be 'examples of teaching' to spaces for knowledge construction about the natural world with an appreciation of its complexity and diversity. We argue that greater consideration of linguistic practices is necessary and urgent in a time where humanity is stepping into profound unknowns with no set boundaries between knowledge and action (Harding 2008). Helping students to acquire a more dynamic and critical view of science is the first step towards promoting a more democratic society constituted by citizens who are not only prepared at the level of content and notions but who can creatively frame contexts and produce new interactional genres for a more equitable and sustainable society.

References

Abd-El-Khalick, F., and N.G. Lederman. 2000. Improving science teachers' conceptions of the nature of science: A critical review of the literature. *International Journal of Science Education* 22: 665–701.

Aikenhead, G.S. 1996. Science education: Border crossing into the subculture of science. *Studies in Science Education* 27: 1–52.

Aikenhead, G. 2004. Science communication with the public: A cross-cultural event. In *Communication in theory and practice*, ed. S.M. Stocklmayer, M.M. Gore, and C. Bryant, 23–47. Dordrecht: Kluwer Academic.

Aikenhead, G., and A.G. Ryan. 1992. The development of a new instrument: "Views on Science, Technology and Society" (VOSTS). *Science Education* 76: 477–91.

Apostolou, A., and V. Koulaidis. 2010. Epistemology and science education: A study of epistemological views of teachers. *Research in Science and Technological Education* 28, no. 2: 149–66.

Brandt, C.B. 2008. Scientific discourse in the academy: A case study of an American Indian undergraduate. *Science Education* 92, no. 5: 826–47.

British Council. 2010. Science and Citizenship. http://www.britishcouncil.org/talkingscience-public-engagement-science-and-citizenship.htm (accessed May 28, 2012).

Camino, E., and M. Dodman. 2009. Language and science. In *Science, society and sustainability: Education and empowerment for an uncertain world*, ed. D. Gray, L. Colucci-Gray, and E. Camino, 71–98. New York: Routledge.

Cole, A.L., and G.J. Knowles. 2010. Arts-informed research. In *Handbook of the arts in qualitative research*, ed. A. Cole and J.G. Knowles, 55–71. London: Sage.

Colucci-Gray, L., E. Camino, G. Barbiero, and D. Gray. 2006. From scientific literacy to sustainability literacy: An ecological framework for education. *Science Education* 90, no. 2: 227–52.

Education Scotland. 2012. What is curriculum for excellence? http://www.educationscotland.gov.uk/thecurriculum/whatiscurriculumforexcellence/index.asp (accessed May 28, 2012).

Engeström, Y. 1987. *Learning by expanding: An activity-theoretical approach to developmental research*. Helsinki: Orienta-Konsultit.

Expert Group Report. 2008. *Science and governance: Taking European knowledge society seriously*. IPOC: Italian Paths of Culture. http://ec.europa.eu/research/science-society/document_library/pdf_06/european-knowledge-society_en.pdf (accessed May 28, 2012).

Fox Keller, H. 2000. *The century of the gene*. Cambridge, MA: Harvard University Press.

Furman, M., and A.C. Barton. 2006. Capturing urban student voices in the creation of a science mini-documentary. *Journal of Research in Science Teaching* 43: 667–94.

Goffman, E. 1956. *The presentation of self in everyday life*. London: Penguin.

Gray, D.S., and T.G.K. Bryce. 2006. Socio-scientific issues in science education: Implications for the professional development of teachers. *Cambridge Journal of Education* 36, no. 2: 171–92.

Gould, S.J. 1998. Foreword. In *Five kingdoms: an illustrated guide to the phyla of life on earth*, ed. L. Margulis and K.V. Schwartz. New York: W.H. Freeman.

Hammersley, M. 2006. Ethnography: Problems and prospects. *Ethnography and Education* 1, no. 1: 3–14.

Harding, S. 1994. Is science multicultural? Challenges, resources, opportunities, uncertainties. *Configurations* 2, no. 2: 310–30.
Harding, S. 2008. *Sciences from below: Feminisms, postcolonialities and modernities*. London: Duke University Press.
Hood, J.C. 2007. Orthodoxy v. power: The defining traits of grounded theory. In *The Sage handbook of grounded theory*, ed. A. Bryant and K. Charmaz, 151–64. London: Sage.
Jasanoff, S. 2007. Technologies of humility. *Nature* 450: 33–4.
Jenkins, E.W. 2000. Constructivism in school science education: Powerful model or the most dangerous intellectual tendency? *Science and Education* 9: 599–610.
Kats, J., and J.T. Csordas. 2003. Phenomenological ethnography in sociology and anthropology. *Ethnography* 4, no. 3: 275–88.
Kramsh, C. 1998. *Language and culture*. Oxford: Oxford University Press.
Krugly-Smolka, E. 1996. Scientific culture, multiculturalism and the science classroom. *Science & Education* 5: 21–9.
Kelly, K. 2008. *The expansion of our ignorance*. http://www.kk.org/thetechnium/archives/2008/10/the_expansion_o.php (accessed May 28, 2012).
Latour, B. 2005. *Reassembling the social: An introduction to actor-network theory*. Oxford: Oxford University Press.
Lee, O. 2002. Equity for culturally and linguistically diverse students in science education: A research agenda. *Teachers College Record* 105, no. 3: 465–89.
Levin, M., and D. Greenwood. 2001. Pragmatic action research and the struggle to transform universities into learning communities. In *Handbook of action research*, ed. P. Reason and H. Bradbury, 103–14. London: Sage.
Longino, H.E. 2002. *The fate of knowledge*. Princeton: Princeton University Press.
Losh, S.C., R. Wilke, and M. Pop. 2007. Some methodological issues with 'Draw a scientist tests' among young children. *International Journal of Science Education* 30, no. 6: 773–92.
Mason, C.L., J.B. Kahle, and A.L. Gardener. 1991. 'Draw-a scientist' test: Future implications. *School Science and Mathematics* 91, no. 50: 193–8.
Matthews, M. 2000. *Time for science education: How teaching the history and philosophy of pendulum motion can contribute to science literacy*. Dordrecht: Kluwer Academic.
Millar, R. 2002. Towards a science curriculum for public understanding. In *Teaching science in secondary school*, ed. S. Amos and R. Boohan, 113–29. London: Routledge.
Nisbet, M.C., and D.M. Scheufele. 2009. What's next for science communication? Promising directions and lingering distractions. *American Journal of Botany* 96, no. 10: 1–12.
Osborne, J. 2010. Arguing to learn in science: The role of collaborative, critical discourse. *Science* 328, no. 5977: 463–6.
Osborne, J., and J. Dillon. 2008. *Science education in Europe: A report to the Nuffield Foundation*. London: Nuffield Foundation.
Quale, A. 2002. The role of metaphor in scientific epistemology: A constructivist perspective and consequences for science education. *Science and Education* 11: 443–57.
Rampton, B. 2007. Neo-Hymesian linguistic ethnography in the United Kingdom. *Journal of Sociolinguistics* 11, no. 5: 584–607.
Ravetz, J. 2003. Models as metaphors. In *Public participation in sustainability science: A handbook*, ed. B. Kasemir, J. Jäger, C.C. Jaeger, and M.T. Gardner. Cambridge: Cambridge University Press.
Söderbaum, P., and J. Brown. 2010. Democratizing economics: Pluralism as a path toward sustainability. *Ecological Economic Reviews, Annals of the New York Academy Science* 1185: 179–95.
Stirling, A. 2010. Keep it complex. *Nature* 468: 1029–31.
Takeshita, C. 2010. The IUD in me: On embodying feminist technoscience studies. *Science as Culture* 19, no. 1: 37–60.
Volk, T. 2003. *Gaia's body*. Cambridge, MA: MIT University Press.
Wieser, B., and S. Karner. 2010. Deliberating genome research: Discursive strategies and performative roles. *Science as Culture* 19, no. 3: 327–49.
Wilson, K. 1992. Thinking about the ethics of fieldwork. In *Fieldwork in developing countries*, ed. S. Devereux and J. Hoddinot. New York: Lynne Rienner.

Appendix 1. Overview of course's activities and data collection points (emboldened).

Course week	Key event/content
1	Introductions (autograph hunt)
	Overview of course and introduction to activity theory
2	**Draw-a scientist test (DAST)***
	Views on science-technology society (VOST)
	Darwin's workshop: science and literature
3	**Collation of VOST results**
	Debate on the role of art in science
	Nature of science and 'measuring' activity
4	Language of science: connections between 'evidence', culture and context:
	Photosynthesis activity
5	Language and science
	Museum visit in preparation for week 6
6	Group presentations on museum visit (formative assessment, 10%)
	Reflective essay task discussed with students
7	Classification process
8	Classification as reflection of human thought: readings from A. Sen
	Chains and webs
	Reflective essay submitted (formative assessment, 20%)
	Easter vacation
9	Communicating science
	Marketing a drug activity
	Communication and pesticides
10	Group presentations on the topic of pesticides (not assessed)
	Final assignment task discussed with students
11	Border crossing and the role of the educator
	End of course evaluations
12	Individual tutorials available on request
12 + 2	Final assignment (summative assessment, 70%)

Appendix 2. Student evaluation questionnaire.

Socio-cultural perspectives on science

Some thoughts from the students...
How did I live the experience of this course? *Please use some adjectives of your choice*
How did I learn through the course?

	A lot	ok	So and so	Nothing at all
Inputs from tutors in the form of lectures				
Discussion in pairs				
Plenary discussions				
Practical activities in class				
Mini-research activities in groups				
Readings				
Personal reflections				
Formative assessment				

The best opportunity(ies) for learning for me was...
Something I was surprised by:
Something I was disappointed by:
An idea I would like to develop further:
A suggestion I would like to make for the future:

Evolution education in policy and practice: an ethnographic perspective

David E. Long

Department of Middle, Secondary, Reading, and Deaf Education, Valdosta State University, Valdosta, GA, USA

Evolution education in the US is conducted unevenly, or in cases is absent. Showing the strength of ethnography as a means of deeper explication in science education, this article explores the interactions of policy and practice in evolution education. Discussing vignettes from a larger ethnographic study, Creationist rationales and practices within the classroom are explored, with implications for science education policy and practice.

Introduction

Evolution education in the US science classrooms is uneven or at worst substantively absent. Against the intent of educational policy ideals such as the American Association for the Advancement of Science's (AAAS) *Science For All Americans* (Ahlgren and Rutherford 1990), science teachers often qualify or omit evolution from the explicit curriculum. As a part of these AAAS ideals, students are to be educated to exhibit scientific 'habits of mind'. These habits of mind, as AAAS sees them, call for an epistemological perspective, respective of varying ways of knowing, but ultimately see appropriate knowledge as grounded in a naturalistic perspective.

Usually disconnected from policy makers, science educators of varying epistemological commitments staff US schools. Local cultural values and religious traditions may work against developing these AAAS ideals. As one measure of the outcomes of this, Lerner's (2000) state-by-state analysis of science education standards shows that evolution is often omitted or downplayed at the state level. While the day-to-day practices of scientific research press on, American Creationist efforts continue to attempt to tinker with what constitutes science. This was most recently underscored in the Dover Pennsylvania Intelligent Design Trial of 2005.[1]

The majority of educational research designed to tackle anti-evolution sentiment in the US has focused on correcting 'misconceptions' people have, regarding the content of science. Arguably rutted in a kind of naive positivist view of 'the real', as Shumar (2009) has described, this large body of educational research – focused on imparting a better understanding of evolution has yet to argue compellingly as to why, against the facticity of evolution, many people *know* otherwise. This knowing, borne of contingency and culturally imbued ontology, can be described. People live in places situated in the world and know many things apart from science, regardless of those within science education who wish to control and centre all knowledge stemming from a view of science as the master epistemology. At the same time, epistemologies that conclude evolution has not taken place, like all epistemology, can be deconstructed. Doing this in educational practice, with students, requires work far exceeding the constructs of epistemology. Not doing so leaves the field of science education with simple descriptions of scientific illiteracy, a view of cultural deficiency. In the American context, given the large minority that practice some form of Christian fundamentalism, the work of advocating scientific understanding is made harder. With the American practice of separating church and state matters, at least in a loose *de jure* sense, the matter of religious disposition and the matter of state education are to be kept apart. As the status of evolution education makes clear enough, such a distinction is rarely adhered to closely.

Seeing education as a matter of acculturation or gross socialisation to community values, ways of practice and intellectual traditions, how does the concept of evolution fit with those whose community values that stand squarely against it? For example, the millions of American Evangelical Christians who *know* they have the Truth, the light and the way? To learn about evolution for many requires adjudicating competing messages from school, houses of worship, the media, family and friends. What knowledge counts in this milieu? What is the role of science education within this?

Turning an anthropological lens on education, in a 2007 piece in the *Teacher's College Record*, Varenne asks a helpful question: 'Where do people learn about Creationism and Intelligent Design?' (Varenne 2007, 1562). Although anthropologists such as Toumey have done work with Creationist engineers in the American South (2004, 1994) articulating the ideologies and rationales by which they stand apart from their more intellectually mainstream colleagues, such perspectives have yet to be illustrated in a robust way for students involved in evolution education. This work begins that process. Seeing continued resistance and little change in US attitudes towards evolution over time, this article discusses the strength of ethnography as a framework and approach to dissolve, or at least to begin discussing ontological issues that crop up when science and religion are wielded as epistemological weapons. Demonstrating the interpretive power of ethnography for such issues in science education, I pull three narratives from my larger ethnographic work (Long 2011) to explicate the commitments of culture which preclude some from apprehending evolutionary theory as a simple, matter-of-fact conclusion from scientific reason. As part of an ethnographic study of college students taking their required biology coursework, this article presents these vignettes as points indicative of the 'difficult deliberations', as Varenne terms it, involved in evolution education.

To discuss these issues, I first give a short review of American science literacy ideals, and recent critical discussions as to their perhaps limited potential for having broad impact within the larger cultural milieu. Explicitly joining the methodological focus of this special issue of *Ethnography and Education*, I next discuss three exemplars pulled from my ethnographic work with college students. These student and teacher stories illustrate clear cases where the variable influences of formal and informal learning, once articulated within a theoretical framework of cultural commitment, account for and present difficult scenarios for getting evolution education done. Secondly, these stories bring forward a critical issue for the practice of science education. The often claimed and sought-after value-neutrality of the 'scientific objective' view is called into question by Creationists who rightly and instructively point out that AAAS-styled scientific practice is as much part of a social agenda (social justice) as Creationists themselves are. Although this is something many like-minded AAAS policy writers might like left unarticulated, recent moves in 'Intelligent Design' movement appear set to vigorously pursue this line of argument. As Fish (2005) has written about, one of the unintended consequences of post-modernism's success within academe is the dramatic tolerance and traction of a 'teach all the theories' accommodation by teachers within classrooms.

The analysis of this article's thematic foci draws on the categorical framework of Wuthnow's (2005) sociology of religion and the conceptual tools of Swidler's (1986) 'strategies of action' to articulate the limitations that US educational structures place on a more robust dialectic towards evolution. Although already well-known religious disposition usually dictates one's attitude towards evolution. At the same time, from the corpus of my data, few students or faculty articulated a rich understanding of their own religious identity's theology, and almost none could articulate interfaith distinctions. This is instructive as resistance to evolution is almost always grounded in religious terms. As educational discourses towards evolution have moved towards a serious consideration of worldview as an interpretive framework for science education (Matthews et al. 2009; Reiss 2009), this work points towards lacking religious education as a culprit in American anti-evolution sentiment.

Against policy intentions, this research illustrates the 'practical reason' (Bourdieu 1998) involved in student lives as they resist evolutionary theory. People may have quite 'reasoned' rationales based in cultural logics against the ends of scientific orthodoxy. At the same time, due to the risk of producing contention, many American educators downplay evolution as not to risk offense. In this way, this research turns the lens back on Varenne's (1986) earlier work on American 'friendliness'. Part of the unstated negotiation of evolution education for many involves a careful avoidance. Evolution education, with all of its potential contention, seems better described for its deliberative absence rather than presence. Science education policy that does not consider such phenomena, and the rationales behind them, is a poorly thought out policy.

Method and direction

An ethnographic study of 31 college students working through their introductory college biology course was conducted in the spring and summer of 2009. Grounded in a phenomenological perspective, in interviews I asked the students and faculty to articulate elements of their worldview, and how it related to the concept of evolution. Phenomenology is important to understanding resistance to evolution as I aim to make sense of how 'how [students] perceive it, describe it, feel about it, judge it, remember it, make sense of it, and talk about it with others' (Patton 2002). Given that the downplaying of evolution in education has resulted in social contention being 'covered-up', I draw on Dreyfus's (1991, xvii) distinction about phenomenology that 'the phenomenologist has to begin by describing and working through the everyday cover-up to arrive at the phenomenon that is being concealed'. As such, my general research paradigm proceeds along the lines of Flyvbjerg's (2006) *phronetic* social science. I am interested in what the extreme case regarding resistance to evolution can teach us about the general.

To approximate a normative educational experience by which I might say something useful to regions of the US that struggle with anti-evolutionary movements, I chose to locate my research at a regional Master's degree granting institution. As Levine (2006) sees it, this would be the most normative case for both the type of institution that graduates the most students, and produces the most future teachers. Mason-Dixon State University sits in the middle of America, in a major river valley. Increasing the chance that evolution might be a point of contention in the university curriculum, Mason-Dixon is reasonably close by the Answers in Genesis Creation Museum, which opened in 2007 to the chagrin of scientists and some science educators. Mason-Dixon, like US universities on whole, enrolls slightly more women than men. While it almost matches US averages for ethnic minority enrolment, this group was almost entirely African American. Religious organisations were prevalent and active in campus programming.

During the spring of 2009, I followed 31 students throughout three differing introductory biology classes, listening to and recording class discourse, interviewing them as they thought about evolutionary theory, and participating in campus religious programming that contested evolution. Biology faculty, other university faculty, campus religious group leaders and students' past teachers were also interviewed. In these interviews students, faculty and past secondary school biology teachers discussed their life histories with science and religion and how these have interrelated during their lives. Compiling their narratives, students recall how and where the controversial status of evolution in American culture became clear to them, and how this was cultivated in the classroom and church, if at all. Describing their feelings regarding evolution, Creationist students express existential anxiety, and flee from contemplating the topic. By doing this, I started at what I consider the likely end of the road for evolution in one's formal education, working my way back through student life histories, school experiences and community contexts on the topic. While this paper focuses on specific ethnographic data's strength for unpacking understanding in science, the full ethnographic project appears in Long (2011).

With an eye towards looking at the efficacy of evolution education as currently done, as this research demonstrates, few American students come to college with a prior education in evolutionary theory anywhere close to that which policy documents such as the American Association for the Advancement of Science's *Project 2061* advocates (AAAS 2001, 1993). A large cadre of part-time and adjunct faculty, symptomatic of an increasingly fragmented academic work-life, often teach the non-majors science classes. Many of these faculty have shied away from the topic of evolution—fearful and tired of student challenge, and unprotected by the traditional intellectual security of tenure. Tenured faculty describe a phenomenon of 'turning-off' by students when the topic is discussed. Classroom rhetoric is often carefully shaped to avoid conflict.

Students, teachers and the scientific ideal the AAAS intends

Since the mid 1980s, the AAAS has sought to align a common science curricular roadmap for American science education, providing *Benchmarks* (1993), and an *Atlas of Science* Literacy (2001) to read along the way. At the same time, some such as Shamos (1995) have rightly expressed scepticism at the plausibility of achieving these overarching goals. More recently, Hodson (2003) and Reiss (2009) have varyingly outlined warrants by which science should be a tool for social action. Brandt and Hammond (2004) have also refined such views within the anthropological frame to mobilize science as a tool of social justice. This paper continues in this same vein in general as seeing science as an agentic tool for individual and/or community good, but remaining mindful of critique from those disinclined to reframe science in this way.

With this in mind, I aim to illustrate the difficulty in negotiating evolution education across ideologies. As Hodson points out regarding any radical restructuring of science education's explicit ends:

> Making the kind of changes to the curriculum [as] advocated . . . will not be easy. Much that I have suggested is likely to be disturbing to science teachers, severely testing both their competence and confidence. Traditionally, science education has dealt with established and secure knowledge, while contested knowledge, multiple solutions, controversy and ethics have been excluded. (664)

As my thematic exposition will show, moves such as those that Hodson proposes are almost certainly unenforceable. Creationist ontology rests on a foundation of unquestionable, final interpretive authority. While scepticism and humility in the face of science's open possibility are virtues in science, they are not hallmarks of Creationist thought. When we move an ideal of science as tool for social justice as a politically active educational programme against contentious curricular content such as evolution, the problems become clearer. Resistance to evolution comes along with a gestalt package of ideological commitments, which will almost certainly work against the ideals I or other like-minded writers may set. Science educationalists, and for that matter educationalists in general, have no way of ensuring the dispositional allegiance of their colleagues regarding matters which switch from the atomistic science-as-object that it has historically worked to discuss towards a more politically active social programme. As Damon (2005) and Villegas (2007) demonstrate, this enters the highly contentious area of teacher disposition and the politically charged terms with which this debate is armed. As it stands currently regarding the creation of a workforce of teachers as social justice advocates, there seems to be no lawful or enforceable way to assure such moves on a large scale.

When considering larger ideological ends within gross projects for science literacy such as social justice, one should not be surprised to also find ideological resistance from Creationists. For example, AAAS documents, such as the *Atlas*, *Science for All Americans* and *Benchmarks for Science Literacy*, have foundational premises such as: science literacy is needed to 'prepare people to lead personally fulfilling and responsible lives . . . [which] should help to develop the understandings and habits of mind they need to become compassionate human beings able to think for themselves and face the world head-on' (1993, v). This ideological claim for educating such an individual in the world is linked to preparing citizens for national concerns such as:

> [equipping] them also to participate thoughtfully with fellow citizens in building an protecting a society that is open, decent and vital. America's future—its ability to create a truly just society, to sustain its economic vitality, and to remain secure in a world torn by hostilities—depends more than ever on the character and quality of the education that the nation provides for its children (vi).

Turning this towards controversies in American evolution education, Creationists have historically presented 'scientific evidence' that supposedly affirms Biblical truth of one sort or another towards other ends. Although Ham (1987) warns of the 'dangers of evolution' leading to a society rife with 'evils' such as 'homosexuality and abortion', more subtle, but still ultimately Creationist views such as Dembski (2004) or Behe (1996) present the concept of 'irreducible complexity' as completely overturning evolution as a scientific paradigm. As Forrest and Gross' (2004) uncovering of the Discovery Institute 'Wedge Document'[2] has shown, this type of argument is not only non-scientific, but is attached to a deep-seated motivation to affirm and grow a pan-Christian theocratic nation. Michael Behe, ID proponent and *Kitzmiller* expert witness attests what's at stake:

> Christians live in the world with non-Christians. We want to share the Good News with those who have not yet grasped it, and to defend the faith against attacks. Materialism is both a weapon that many antagonists use against Christianity and a stumbling block to some who would otherwise enter the church. ... Although Christianity can live with a world where physical evidence of God's action is hard to discern, materialism has a tough time with a universe that reeks of design. (Behe 1996)

Between the extremes of AAAS's ideally scientifically literate citizen and the unabashed theocratic agenda of Creationism, what does the landscape of evolution education, in practice look like? Just how much of the teaching force wants to get evolution education done? The literature of science education regarding the attitudes towards evolution by teachers is disheartening. As a summary, the works of Eglin (1984), Jorstad (2002), Kyzer (2004), Moore (2007) and Zimmerman (1987) are good insights into the American teaching profession's commitment to perform high-quality evolution education. As shorthand towards this research, the majority of American teachers – likely neither religious zealots nor scientific experts of the highest caliber – are glossing, avoiding or omitting the concept from their teaching. Whereas this prior work has tended to be dominated by decontextualised survey research, the following ethnographic vignettes point to the richness available through humanising evolution's meaning for those most disposed to reject it, along with the mundane and mostly detached view towards evolution that most have. These stories elucidate three key themes with which I will discuss the importance of ethnography for further the aims of science education.

Demonstrating the need for an ethnographic lens on evolution education

Besides the policy ideals of organisations like the AAAS, teachers of various ideological dispositions staff American classrooms. Like American society in general, teachers on whole are fairly religious. As Kimball (2009) has found, education majors are in fact the most religious of all college majors in the US. Layered upon this, the Pew (2008) religious landscape survey finds that although Americans are slowly becoming less religious, of those who have faith, Evangelical Christianity continues to grow, with fundamentalist Evangelical Christianity[3] the fastest growing percentage of believers. The following vignettes demonstrate how some of these people, both teachers and students, negotiate learning and working within this system to complicate evolution education.

Creationists experiencing existential anxiety

Preliminary interviews with Creationist students revealed the emotional stakes and perceived marginalisation that these students sometimes faced. Students identified themselves categorically as Creationists (9 of 31 students) in interviews by the nature of their critique of evolutionary theory. As Ruth, an advanced student close to graduating describes, her past did not set her up with a receptive view towards evolutionary science.

> My parents were missionaries, and I've grown up Christian my whole life, so like, we didn't have that debate in school like, '*should* we teach evolution?' I personally don't believe in evolution, so it was kind of a shock to me when I had just come to college. I took intro bio in my freshman year... and I went through that whole 'what do I believe... is it my parents or is it other things'... I think that was my first contact with it. So, [evolution] was like huge in class, and [the professor] presented it as there could be no other explanation... and, I mean I remember him saying that if you believe in God creating the earth, then pretty much you're an idiot. And he obviously didn't use those words, but that's pretty much what he said. And I remember thinking... *What is he talking about?*[4]

Although challenged epistemologically by the orthodoxy of evolutionary science, Ruth nonetheless was not challenged pedagogically.

> I take those really big classes, because it's really easy to excel in those huge classes. I mean, I got like a hundred on every test. You have to be an idiot pretty much not to. If you just sit, and you listen to what they're saying, and you know how to take tests, it's very easy to do well in those classes.

Tyson, a biology major planning on becoming a dentist, was perhaps one step closer towards having to adjudicate his Creationist views with the epistemology of science. Having been homeschooled in a Evangelical Christian family, Tyson now ruminates on his Creationist family, church and identity.

> It's a big reason why, in the last year or two years, I really questioned it. My parents are Creationists, but I don't know if I believe this stuff. So I was like going through trying to figure out what I believe. They all say it's scientific, but I'm like, 'Really?' A lot of the reason I'm Christian or a Creationists is because my parents are and my friends are. Social life plays so much into that.

Crucially in all these interviews, after a Creationist student would detail what they saw lacking in evolutionary theory, I asked them to imagine: what if they had to consider that the evidence for evolution really was there—what would it mean to their lives if evolution had irrefutably happened?

> It would be a complete crisis. It would be really tough. I've asked people this. Some people don't believe evolution because the bible says so. I'm like, 'You have to have some rational reason'. People believe the Qur'an and all that stuff. We don't believe that. So then it would be a big crisis. The Bible teaches evolution is not true. If the Bible is not true, then there is no afterlife. I don't know why I should be good in the first place. There's a lot of conflict there. I'd have to try to find some other evidence. If there was no evidence, I'd either give it up, or I'd be too lazy to give it up because I want to keep my friends.

Cindy, a future elementary school teacher, had been raised in the Evangelical Christian tradition. Having been scorned by her church community for having an abortion her first year in college, Cindy moved away to another university, and stands on the other side of Tyson's feared abandonment of friends and social connection. Instructively, evolution now is seen as a possibility.

> I have not been to church in about three years. There were some things in my life that happened . . . and the congregation started talking about it. After that I kept thinking – Well, is the church really here to go to church, or are they here to gossip? So I really haven‘t been back to church. Plus, with me questioning everything and how everything evolved, I just don‘t feel like church is really the right place for me at this time in my life.

Cindy continues to explain her reframed view of the world and the importance of those few on the other side of this shift. The life experience of her fiancée has Cindy questioning the fixity of her worldview. 'He‘s been absolutely everywhere. He has seen various parts of the country, and different cultures, and different types of people. And he has just not convinced that one vast religion should be able to say this is why we are'. When I ask how she feels about the changes in her spiritual life, the emotional investment in religious practice lost haunts Cindy.

> From time to time, whenever I'm not busy, I think about it. I don‘t really like that I don‘t have any kind of religious thought right now. I was so used to going to church and youth group. I think a big reason that I have not picked a church, or attempted to go to church, is I have not wanted to have that discussion with my fiancée. Oh, he's already told me that I could go, but he doesn‘t want me coming back home and talking about it because he don‘t want to talk about it, so . . . (*pause*) and he's my only person here [at college].

Finally, Cindy discusses what she sees as being different in her thinking about evolution now, as opposed to earlier in her life.

> I think it‘s mostly about being scared. I think in my teen years is when I really realized that death was factual. I think just I didn't want to think that there wasn‘t a heaven. I guess I feel that if evolution did happen the way that we were talking, being strictly scientific, I guess it made me scared because then the whole myth of God or whatever, wouldn‘t be true. And it was just be like – say a dog dying – dog dies, she's gone. And I would just be gone. So I just don‘t think I wanted to think about that back then.

Sociologist of religion Robert Wuthnow has written extensively (2005) about the structure of American religious faith, and how these belief systems affect the sociality of lives. Importantly to public receptivity towards evolution, Wuthnow has framed these categories towards the topic of evolution in debates between science and religion (2009). Wuthnow sees American religious practitioners as fitting in three ontological categories of belief and non-belief. Religious exclusivists, who would include almost certainly all Creationists, are those whose religious ontology directs them to see one unified Truth which they have in their possession. Religious inclusivists are those inclined to see many ways to truth, respective of others. The category of non-believers, a small percentage of the US population, fills a third category.

As Wuthnow discusses, religious exclusivists live in tight social circles in which they have little meaningful exposure to other religious interpretations, and make practically no attempt to understand others. In cases such as Cindy, when this interpretive frame becomes broken, one does not supplant it with another more liberal form of theism, one often abandons established connections to religious thought altogether. Prior to finding one's self in such an existential lurch, the anxiety or dread of the conceptual abyss Kierkegaard (1957) and Tillich (1952) worked to articulate, keep students and for that matter adults from considering other options. In short, due to no common cross-societal religious knowledge, Americans are limited in their ability to think more deeply and creatively about the most crucial metaphysical matters of science and religion.

Constructing contention over evolution

The adoption of the ethnographic lens also helps to see cultural constructions in more explicit terms in the rhetorical milieu of education writ large. Evolutionary scientists on whole might not choose to investigate how and why evolution becomes controversial in American society. But how and why is contention over evolution learned? The influence of Creationist churches is easy enough to follow in my other work (Long, 2011), and outside the purview and intent of this short paper. What is compelling from my research is the degree to which school teachers themselves were the first and primary agents for students in these regards.

Across my interviews, I asked students where and when they first became aware that evolution is a controversial topic, if they ever had in their lives. Time and time again, and often including teachers who held meta-dialogues of qualification prior to 'the evolution class', students reported first hearing evolution as controversial from their middle school science teachers. Carly first remembers hearing this in 'middle school biology . . . I guess that's when they first mention it. But they don't really talk about it too much because you're not supposed to teach it supposedly'. When I ask her to clarify exactly who stated this, she confirms 'the science teacher'.

Tanya, who attended a speciality high school that consistently scored in the top ten for academic performance in her state was clear about evolution not being taught. 'We don't talk about evolution . . . or we didn't'. When I asked her to clarify why she hesitated, she explains that the topic might come up in a side discussion regarding the age of fossils or the lineage of certain animal species, but these questions always met curt responses from the teacher who added 'but there's also other theories as to how we came about'. Indicative of the kinds of administrative pressures I heard discussed in a few schools, James learns that evolution was to be avoided at school.

> It was an iffy subject. It wasn't something that they were . . . I guess the principal, or maybe the school board told them not to delve too deep into it. It was kind of a faux-pas/taboo area when I was in school at least. Coach Sims, the teacher . . . when he brought up evolution he was like well, you know, 'but I can't go into it because they told me not to' – like the board [of education] told him not to.

Cindy's perceptual experience in a community dominated by Evangelical Christians described the same kind of ideologically charged pressure. 'The teachers didn't want to go too in depth with it because they were afraid that they were hitting points that the students didn't want to hear due to religion'. Cindy's high school teacher seemed frustrated by the social pressure to not teach about evolution. After showing a short video 'approved' by the administration on evolution, she sarcastically summarizes, 'well . . . we covered it'. Cindy's case was extreme but not unusual. It appears that some school discourses about science, and student's perceptions about them, are simply heavily charged due to the general cultural influence of religious exclusivist worldviews and what these dictate in normative social practice.

Andrea, while not attending a school that eliminated evolution, did have teachers fully aware and practised in how to handle certain 'trouble students'. As one of Andrea's friends alerts her to the impending 'doom' of evolution:

> One of my friends that was in the class before mine with the same teacher had mentioned to the teacher, 'oh well! Andrea isn't going to like this because I just don't believe in it'. So the teacher came up to me before class started and warned me, 'Hey, we're going to be talking about this, but I'm not going to teach it as a fact, so you don't need to worry'. Like really making sure that I wasn't going to cause a big deal. And I'm not the kind of person who would do that. I just kind of sit back to myself.

Contention over evolution, as Andrea discloses, can be easily enough suppressed.

Cindy, like Andrea, who both are preparing to become teachers clearly saw the problems of teaching evolution cleanly in her community. When I ask her how she might achieve this:

> If you want people to learn about it, you're going to have to approach it in a way that they're not going to be offended by it. So like in my hometown (*pauses*) I'm trying to think how you would approach it. I mean, there would be no easy way to approach it, really.

How Creationist teachers can slide by unnoticed

Almost all previous research concerning teacher and student attitudes towards evolutionary theory, as seen in the briefest form above, has been broad-scale survey research. Although steps have been taken towards more ethnographic approaches (see Kyzer and Goldston 2009), there remains precious little ethnographic perspective in a field rife with contention and possible inquiry. What is missing from this research is an articulation of rationales and actual practices that those who downplay evolution in the classroom use and take part in. Switching now from student to teacher narratives, two particularly poignant cases illustrate the depth and difficulty of accounting for Creationist influence within science classes.

It has often been suggested within the US context that social controversy concerning evolution in science could be effectively and instructively taught within the content of a history, social issues or perhaps comparative religion class. One of my college student interviewees had experienced such a class in high school – comparative religions with Mr Woodman. When I met to interview Woodman, he walked me through the content of the class. Two things became clear: he was using his public school class as a vehicle to critique evolutionary theory using a case-by-case comparison of world religions. First:

> What I wanted to do in terms of comparing all, was for the curriculum to be set up as a broad overview. I wanted all the religions underneath that umbrella to always apply the concepts of Intelligent Design versus evolution. That way the kids could say, 'Okay, we study this. We study Catholicism. Now where does that fall under here? How would Catholicism discount evolution?' And then I wanted them to figure out or debate – 'now that you know a little background about Catholicism' – or we would do Judaism or Hindu, or whatever. Then I would have to break them up and go, 'Okay, now you're gonna take on the role of – you are now a believer of this. And *you* are a believer, we talked about, of evolution. We're gonna debate this'.

Secondly he also (unsurprisingly) was an Evangelical Christian.

> The teachings of my religion would definitely convince me that evolution is not possible. The religious teachings I have would say that if man came from a primate, that the Intelligent Design piece of it, the spiritual side of religious teaching, it's possible that primates started first, but we didn't evolve from primates.

Ms Garcia proved an even more pointed example. Having only the slightest inclination that Garcia was a Creationist prior to my interview the following institutional dynamics were eye-opening to say the least. Upon her hire, as Garcia explained to me, a litmus test question was posed to see whether she was willing to teach evolution in this public school.

> In my interview they said, 'how do you feel about teaching evolution?', and I said, 'Well, if I have to teach it I have to teach it. I'm okay with that you know – I've taught it before', and they said, 'Well, we don't teach it here'. I just said, 'Okay, well then what's the point of asking then?', you know, and then the principal just kind of laughed. She said, 'I just want to know your opinion on it'.

As Garcia explained to me, the school community is heavily influenced by local Evangelical church congregations that prompt students to demand alternative assignments if the school presents knowledge claims 'against their religion'. Instructively, this scenario proved no major ethical or professional conundrum for Garcia.

> I'm a Christian. I believe that . . . you know . . . we came into existence in that way. I don't believe that we evolved in any way. Well like you know, in the teacher's lounge talking with my friends it's highly controversial. We have a very strong Christian network here at school.

Most insightfully as I noticed during our interview that the origins of the universe was the class topic that day, Garcia shared with me an example of how she treated this.

> The students were like . . . 'Well how do you know that a big bang created it?' and I said, 'I don't, but I believe that somebody else did and he's the Almighty Lord' and that's okay. It's my flair as a teacher – when I'm teaching then it's my flair as a teacher to express my personal opinions because I'm a person. I teach it because the state says I

> need to, but I could also express my opinions. I'm going to test them on the Big Bang Theory, but I also can express other things.

Evolution, religious reflexivity and new avenues in science education

My intent in presenting these narrative vignettes was to illustrate ways of understanding the process of evolution education exemplifying the difficult and political and rhetorical nature of science education. As Apple (1979) has put it plainly, 'it is naïve to think of the school curriculum as neutral knowledge'. But working students and teachers to a discursive space where the preceding understandings of science and religion are made more explicit, requires a more aggressive approach to historicising, philosophically grounding and directly addressing issues of social power in knowledge construction. In almost each case, students and teachers I encountered that held Creationist views spoke simply and synecdochically of Christianity. They possessed *the* one true way.

In my estimation, Swidler's (2001) cultural toolbox metaphor, in the tradition of Bourdieu (1977), gives us a fairly simple means by which to directly address the limited intellectual repertoire of Creationism. Creationists, as seen in Wuthnow's framework, are usually bereft of a dexterous vocabulary with which to speak about matters of faith—as they have learned at home and in church that they possess a final Truth. If the skill to speak about matters of faith and science interpretively can be seen as a form of cultural capital, Creationists not only did not possess it – they seemed to reject it as a form of fungible currency. Perhaps ironically, a few voices within the 'New Atheists' movement, namely Hitchens (2007) and Dennett (2006), have rightly suggested that the cure for, or politically active stance against religious exclusivism may in fact be to support the state teaching of religion in the US. In this way, a breadth of religious epistemological commitments might enter the conceptual vocabulary for more students. At the same time, one might predict that religious exclusivists might be the first to resist such efforts, fearing a loss of control of the authority of the message.

Reflecting the limited way many students and teachers above conceived of both science and religious faith, Swidler's (2001) theorising of culture as shaping discrete pathways of action goes a long way to emphasize this.

> Think of [someone] as a hiker ascending a mountain, with culture as a description of the path she follows. The mountain's topography will certainly affect her route. She will pay attention to a boulder she must cross or go around, to steep or flat places, to openings in the trees. But other features on the mountain that do not directly affect her climb may be irrelevant. She may misconstrue the larger shape of the mountain, yet well describe her own path. (132)

In Swidler's conceptual terrain, abandonment of a Creationist commitment is to abandon a familiar and socially significant strategy of being.

> People do not readily take advantage of new structural opportunities which would require them to abandon established ways of life. This is not because they cling to cultural values, but because they are reluctant to abandon familiar strategies of action for which they have the cultural equipment. (281)

As this special issue has addressed, some science education researchers see science as an agentic tool – one that can be used to rectify and/or subvert forms of social inequity. At the same time, religious exclusivists work diligently to tamp down multiple perspectives of knowledge in favour of one final immanent frame. Ethnography provides a methodological avenue for both; humanely articulating the competing rationales of seemingly incommensurable worldviews; and for informing educational policy that works to structure an education system in service of the fullness of democratic education filled with possibility. Pointing new paths forward for evolution and science education generally will almost certainly require showing myriad ways that scientific and religious paths have been paved. Working through this in a practical sense – sitting religious exclusivists, religious inclusivists and secularist down to constructive dialogue – is work yet to come.

Notes

1. The *Kitzmiller v. Dover* ruling in US Federal Court delineated that 'Intelligent Design' theory, as science, has no empirically founded evidentiary content.
2. The Wedge Document, made famous in *Kitzmiller v. Dover*, was an internal strategic plan of the Discovery Institute oriented towards infiltrating public school institutions to slowly establish Biblical authority in science classrooms.
3. I use fundamentalist in this instance to categorize any Christian denomination who holds Biblical scripture as a final, literal, authoritative Truth, not open to interpretation. This category has been estimated to be 25–30% of the US Population, or 70–80 million Americans (Institute for the Study of American Evangelicals 2011).
4. Student and teacher quotes discussed within this paper appear in full ethnographic context in Long (2011).

References

Ahlgren, A., and F.J. Rutherford. 1990. *Science for all Americans.* New York: Oxford Univ. Press.

American Association for the Advancement of Science, Project 2061. 1993 *Benchmarks for science literacy.* New York: Oxford Univ. Press.

American Association for the Advancement of Science, Project 2061. 2001. *Atlas of science literacy.* Washington, DC: American Association for the Advancement of Science.

Apple, M.W. 1979. *Ideology and curriculum.* Routledge education books. London: Routledge & K. Paul.

Behe, M.J. 1996. *Darwin's black box: The biochemical challenge to evolution.* New York: Free Press.

Bourdieu, P. 1977. *Outline of a theory of practice.* Cambridge studies in social anthropology, 16. Cambridge: Cambridge Univ. Press.

Bourdieu, P. 1998. *Practical reason: On the theory of action.* Stanford, CA: Stanford Univ. Press.

Brandt, C., and L. Hammond. 2004. Science and cultural process: Defining an anthropological approach to science education. *Studies in Science Education* 40: 1.

Damon, W. 2005. Foreword: Personality test: The dispositional dispute in teacher preparation today, and what to do about it. *Arresting Insights in Education* 2, no. 3: 1–6.

Dembski, W.A. 2004. *The design revolution: Answering the toughest questions about intelligent design.* Downers Grove, IL: InterVarsity Press.

Dennett, D.C. 2006. *Breaking the spell: Religion as a natural phenomenon.* New York: Viking.

Dreyfus, H.L. 1991. *Being-in-the-world: A commentary on Heidegger's Being and time, division I.* Cambridge, MA: MIT Press.

Eglin, P.G. 1984. Creationism vs. evolution: A study of the opinions of Georgia science teachers. PhD thesis, Georgia State University.

Fish, S. 2005. Briefings – Academic cross dressing - How intelligent design gets its arguments from the left. *Harper's.* 70.

Flyvbjerg, B. 2006. Five misunderstandings about case-study research. *Qualitative Inquiry* 12, no. 2: 219–45.

Forrest, B., and P.R. Gross. 2004. *Creationism's Trojan horse: The wedge of intelligent design.* Oxford: Oxford Univ. Press.

Ham, K. 1987. *The lie: Evolution.* Colorado Springs: Master Books.

Hitchens, C. 2007. *God is not great: How religion poisons everything.* New York: Twelve.

Hodson, D. 2003. Time for action: Science education for an alternative future. *International Journal of Science Education* 25, no. 6: 645–70.

Institute for the Study of American Evangelicals. 2011. How many Evangelicals are there? http://isae.wheaton.edu/defining-evangelicalism/how-many-evangelicals-are-there/.

Jorstad, S. 2002. An analysis of factors influencing the teaching of evolution and creation by Arizona high school biology teachers. PhD thesis, University of Arizona.

Kimball, M.S. 2009. *Empirics on the origins of preferences: The case of college major and religiosity.* Cambridge, MA: National Bureau of Economic Research.

Kierkegaard, S. 1957. *The concept of dread.* Princeton: Princeton Univ. Press.

Kyzer, P.M. 2004. Three Southern high school biology teachers' perspectives on teaching evolution: Sociocultural influences. PhD Thesis, University of Alabama.

Kyzer, P., and M.J. Goldston. 2009. Teaching evolution: Narratives with a view from three southern biology teachers in the USA. *Journal of Research in Science Teaching* 46, no. 7: 762–90.

Lerner, L.S. 2000. *Good science, bad science: Teaching evolution in the states.* Washington, DC: Thomas B. Fordham Foundation.

Levine, A. 2006. *Educating school teachers.* Washington, DC: Education Schools Project.

Long, D.E. 2011. *Evolution and religion in American education: An ethnography.* Dordrecht: Springer.

Matthews, M.R. 2009. *Science, worldviews and education.* Dordrecht: Springer.

Moore, R. 2007. The differing perceptions of teachers & students regarding teachers' emphasis on evolution in high school biology classrooms. *American Biology Teacher* 69, no. 5: 268–72.

Patton, M.Q. 2002. *Qualitative research and evaluation methods.* Thousand Oaks, CA: Sage.

Pew Foundation. 2008. U.S. religious landscape survey. http://religions.pewforum.org/pdf/report-religious-landscape-study-full.pdf (Accessed February 25, 2008).

Reiss, M.J. 2009. Imagining the world: The significance of religious worldviews for science education. *Science & Education* 18, nos. 6–7: 783–96.

Shamos, M.H. 1995. *The myth of scientific literacy.* New Brunswick, NJ: Rutgers Univ. Press.

Shumar, W. 2009. *Caught in the contradiction of method: A meta-reflection.* Paper presented at the 2009 Oxford Ethnography Conference, September, in Oxford, UK.

Swidler, A. 1986. Culture in action: Symbols and strategies. *American Sociological Review* 51, no. 2: 273–86.

Swidler, A. 2001. *Talk of love: How culture matters.* Chicago: University of Chicago Press.

Tillich, P. 1952. *The courage to be.* New Haven: Yale Univ. Press.

Toumey, C.P. 1994. *God's own scientists: Creationists in a secular world.* New Brunswick, NJ: Rutgers Univ. Press.

Toumey, C.P. 2004. Introduction to Coleman, S., & Carlin, L. (2004). *The cultures of creationism: Anti-evolution in English-speaking countries.* Aldershot, Hants, England: Ashgate.

Varenne, H. 1986. *Symbolizing America.* Lincoln: University of Nebraska Press.

Varenne, H. 2007. Difficult collective deliberations: Anthropological notes toward a theory of education. *Teachers College Record* 109, no. 7: 1559–88.

Villegas, A. 2007. Dispositions in teacher education. *Journal of Teacher Education* 58, no. 5: 370–80.

Wuthnow, R. 2005. *America and the challenges of religious diversity.* Princeton, NJ: Princeton Univ. Press.

Wuthnow, R. 2009. No contradictions here: Science religion and the culture or all reasonable possibilities. In *The religion and science debate: Why does it continue?*, ed. H.W. Attridge and R.L. Numbers, 155–78. New Haven: Yale University Press.

Zimmerman, M. 1987. The evolution-creation controversy: Opinions of Ohio high school biology teachers. *Ohio Journal of Science* 87, no. 4: 115–25.

Science teaching: a dilemmatic approach

Anna Traianou

Department of Educational Studies, Goldsmiths, University of London, New Cross, London, SE14 6NW, UK

In this paper, I examine the nature of primary science expertise using an ethnographic and sociocultural approach and a theoretical analysis that conceptualises educational practice in terms of the resolution of dilemmas. Using data from an in-depth investigation of the perspective and practice of a single teacher, I discuss some of the *dilemmas* faced by this expert science practitioner as she tries to find the most successful ways for facilitating children's learning. The dilemmas I discuss here describe tensions around how scientific knowledge should be presented to children, how children learn science and modes of pedagogy. I show that while some of these dilemmas arise from a conflict between the teacher's own ideals and the external requirements placed upon her, others reflect important tensions within her own values and commitments. I argue that a language of dilemmas allows us to understand pedagogic practice as *dynamic* in character, as frequently involving difficult choices, in which ideals are traded off against one another. This dilemmatic perspective contrasts sharply both with those views of primary science teaching that treat it as the authoritative transmission of basic skills or facts and those that treat it as the acting out of a coherent ideology.

Introduction

While the study of primary and secondary science education has been a well-established field for many years it is only relatively recently that researchers have begun to employ ethnographic and qualitative approaches to any great extent in their investigations. To a large extent this shift stems from the growing influence of sociocultural perspectives of cognition, which treat knowing and learning as necessarily situated and contingent (see Chaiklin and Lave 1993; Lave and Wenger 1991; Roth 1999; Wertsch 1985). Within this view, the learning of science is seen as a cultural as well as a cognitive activity: a process of *enculturation* into the practices of a particular community. These ideas are becoming more influential in the UK, the US and elsewhere, and they represent a fundamental challenge to dominant views about science education, which tend to conceptualise science as culturally neutral, and view science learning and teaching in *normative* and *abstract* terms (see Hammond and Brandt 2004; Traianou 2006). The close affinity between socio-cultural perspectives and qualitative research is exemplified by the growing number of qualitative studies in the field. For example, McGinnis (2006) used a symbolic interactionist approach to investigate the perceptions and practices of five newly qualified primary science teachers during the first two years of their professional lives (see also McGinnis, Parker, and Graeber 2004). Meanwhile, Roth (2000) has employed autoethnography in studying novice science teachers (see also Roth 2005). There are also a number of studies which illustrate the commitment of many science education researchers to view science as a cultural tool; one that is seen as giving *agency* to all people involved in the learning and teaching of science (for a detailed review of these studies see Hammond and Brandt 2004).

In this paper, I refer to an ethnographic study of primary science expertise, which exemplifies this trend: an in-depth account of the perspective and practice of an expert practitioner (see Traianou 2006). In particular, I explore some of the *dilemmas* faced by this teacher as she tries to find the most successful ways for facilitating the science learning of the children in her class. I use the term 'dilemma' to refer to some of the tensions and complexities she faced in her teaching, and to emphasise the multifaceted ways in which this practitioner thinks and acts (see Adler 2001; Berlak and Berlak 1976, 1981; Hammersley 1977).

In line with a sociocultural perspective, my study argues that the nature of expertise can only be discovered empirically by investigating the perspectives and performances of those who are recognised as experts in their local communities of practice (see Traianou 2006, 2007). In this respect, it challenges much constructivist research into teacher expertise which aims at identifying the extent to which teachers' subject and pedagogical knowledge conforms to, or deviates from, researcher-specified interpretations of what counts as expertise in primary science (e.g. Harlen 1999; Osborne and Simon 1996; Summers, Kruger, and Mant 1997).

However, I did not use the concept of 'dilemmas' in my original analysis, which portrayed teacher expertise as *eclectic* in character, requiring the employment, in a flexible way, of a variety of forms of knowledge (including research knowledge), views of learning and teaching strategies in order to deal successfully with the contingent situations faced in the classroom. One way I sought to conceptualise this was in terms of the notion of *bricolage*, as developed by the anthropologist Claude Lévi-Strauss (1966) and later writers (see for example, Denzin and Lincoln 1994; Nelson, Treichler, and Grossberg 1992). However, more recently, I have come to recognise that the dynamic nature of the pedagogic practice of the expert teacher I studied could be conceptualised in terms of Berlak and Berlak's 'language of dilemmas'. So my aim here is to explore how far this can provide an appropriate language for discussing the complexity of science teaching and the possibilities of change.

The language of teaching dilemmas

There is a well-developed literature on the dilemmatic approach to understanding teaching in general (see for example, Adler 2001; Edwards and Mercer 1987; Hammersley 1977; Jaworki 1994; Lampert 1985; Newman 1997) and some have applied a similar approach more widely (Billig et al. 1988). Much of this literature in the field of education has been inspired by the work of Berlak and Berlak (1976, 1981). They draw on George Herbert Mead's social behaviourism to develop a 'language of dilemmas' that captures the complexity of teaching as well as teachers' ability to inquire into their own practice and change it. Berlak and Berlak's (1981) starting point is the sociological assumption that the primary function of schooling is to ensure the continuity of society, and that the main aim of teachers' work is to transmit to pupils a set of values, views and beliefs that are taken for granted in the relevant society, and which had been imprinted on them through their own schooling and the effects on them of other social institutions. They call this the *person as object* assumption (1981, 24). At the same time, however, they point out that teachers and pupils are not merely passive. They can become self-conscious of the forces and structures that have shaped them, and to some extent they have the capacity to act differently on the basis of their reflections. Berlak and Berlak call this the *human consciousness* assumption and stress that this also means that people can, to some extent, change their world.

Following on from this, the authors argue that while schooling activity cannot be understood without taking into account the subjective experiences of teachers we must also attend to the 'social, cultural, and political forces and structures that are omnipresent in all social situations' (Berlak and Berlak 1981, 24). Indeed, these are a major source of the dilemmas that teachers face. A language of dilemmas, they argued, is 'a means of representing in language the diverse and apparently contradictory patterns of schooling' (Berlak and Berlak 1981, 133) and they propose that it can serve as a set of tools for enquiring into and changing teaching. It can represent and bring to consciousness the contradictions that reside in situations, in individuals and in the larger society, as these are played out within institutional life, and particularly within schooling:

> Each dilemma captures not only the dialectic between alternative views, values, beliefs in persons and in society, but also the dialectic of subject (the acting 'I') and object (the society and culture that are in us and upon us). It does so by formulating in each *act* both the forces which shape teachers' actions (those forces that press toward particular resolutions to a dilemma) and the capacity of teachers not only to select from alternatives, but to create alternatives. (Berlak and Berlak 1981, 124–25, italics in the original)

Berlak and Berlak's language of dilemmas was developed in the context of their ethnographic study of a number of English primary schools in the late 1970s. At this time, English primary education had become widely renowned as an exemplar of progressive forms of practice. Berlak and Berlak selected schools for study that were seen as typical in following the child-centred approach was popular at the time.

In interpreting their data, Berlak and Berlak identified 16 dilemmas, which they divided in three sets – 'control', 'curriculum' and 'societal'. The four control dilemmas capture the tensions regarding the locus and extent of classroom control (e.g. 'teacher vs. child control' 'whole child vs. child as student'). The societal dilemmas refer to contradictions and tensions in schooling patterns related to equality, justice and social relations (e.g. 'equal vs. differential allocation of resources', 'common culture vs. sub-group consciousness'). These are distinguished from the eight 'curriculum' dilemmas, which focus on the transmission of knowledge and ways of knowing and learning (e.g. 'personal vs. public knowledge'; 'knowledge as content vs. knowledge as process') (Berlak and Berlak 1981, 135–75). The authors argue that teaching strategies can be seen as simultaneously 'resolving' multiple dilemmas, and they describe a range of different patterns of resolution.

Berlak and Berlak's study illustrates the potential power of their theoretical approach in 'ordering the flux of classroom life and placing the mundane events of daily school life in the perspectives of culture and time' (1981, 37). At the same time, they are aware of the limitations of language in formulating dilemmas. As they point out, just as language can illuminate, so too it can distort and fragment social activity. For one thing, language categorises and separates consciousness from action and from social content. This separation makes it difficult to talk about schooling as a 'continual process wherein context and consciousness are joined in the acting moment' (see Adler 2001, 53). Berlak and Berlak (1981) emphasise, therefore, that dilemmas can only be treated as 'linguistic constructions that, like lenses, may be used to focus upon the continuous process of persons acting in the social world' (111).

While Berlak and Berlak's study was carried out a long time ago, at a time when educational and social conditions in the UK were very different, I came to realise that their theoretical approach had a lot to offer to the study of primary science expertise. I can do no more than illustrate this here, taking some of their 'control' and 'curriculum' dilemmas and using them to explore some of the complexities in the ways an expert primary science practitioner thinks and acts in the classroom.

The study of primary science expertise

In approaching the study of primary science expertise from a sociocultural perspective, I began by specifically trying to identify a teacher who was recognised in her local school environment as an expert practitioner. Through a number of contacts I was led to Coral (this name is a pseudonym, in order to protect the teacher's anonymity). At the time when I began my research, Coral was in her mid-forties and had been working for seven years as a primary schoolteacher in a large inner city school in an industrial city in Northern England.

The school was located in an area not far from the city centre and served a very diverse local community in terms of class, religion and ethnicity. In order to highlight the diversity of the school's pupil population, the head teacher pointed out to me during our first meeting that pupils in the school spoke 25 different languages. Coral was the class teacher of twenty-eight Year 5 children (10–11 years old), many who had English as their second language. She was also the science coordinator for her school, responsible for the development, overview and monitoring of the school's science programme of study. Science is a core subject in the English National Curriculum, together with Mathematics and English, and must be taught to all year group children (5–11 years old).

An important part of Coral's wider role was to assist colleagues with aspects of their science teaching, especially those related to their understanding of scientific knowledge and procedures, and also to develop assessment and recording procedures which teachers in the school could use, especially in the preparation of children for formal testing. In England, children are assessed in their core school subjects at the age of 7, 11 and 14 years through the use of high stakes tests (Standard Assessment Tests). This testing puts tremendous pressure on teachers, schools and children since the results are made publicly available and influence parental school choice. This kind of testing has been criticised heavily by teachers, researchers and the public, but despite criticisms, Standard Assessment Tasks (SATs) continue to play a crucial role in public evaluations of the quality of school provision. Coral had also published articles in a professional journal of primary science about the ways in which her own teaching was contributing to the development of children's science learning, and at one time she had been a member of the editorial board of that journal. Furthermore, she had presented papers at science education conferences and acted as the science adviser in her Local Education Authority.

I collected data intermittently over a period of two years, initially through classroom observations and interviews with the teacher and some of the children. Coral taught science once a week for about three hours, though she also did other shorter sessions during the week, which usually involved asking children to write up their scientific ideas, explanations and questions. I carried out classroom observations using field notes and audio recording. During group work I joined in with a group of children and observed the ways in which they were getting involved with the practical activities that Coral had set up for them. Early on in the research process, I received an email from Coral in which she included a three-page piece of writing about her work. Coral explained that she had decided to write down her thoughts about what took place during the sessions I had observed, so as to offer me a more detailed account of her teaching, and because she found it stressful to talk about her teaching in school time when there were frequent interruptions. She called this piece of writing her 'diary'. It involved a description of, and justification for, how each of the sessions had been planned, which included a discussion of the aims, a description of what went on, the practical activities the children were asked to engage in, her understanding of how their learning progressed, and above all her concerns about her own teaching in helping children to develop their scientific understanding. She also included in this email examples of children's questions and explanations as an indication of their progression in science. This was the first of many emails I received from her. I found this to be a very important source of data and despite some initial reservations I encouraged Coral to continue to send me similar pieces of writing and tried to establish a collaborative relationship with her (see Traianou 2007).

In the following sections, I include data from an unstructured interview, four email messages and Coral's own writing about the teaching of *forces*, which lasted six weeks. The dilemmas I discuss here can be organised into three groups, concerned with conceptions of knowledge, views about children's learning and the teacher's role. As I was re-analysing the data from the case study, I came to recognise that Berlak and Berlak's work provided a model of the tensions under which the teacher in my study was operating. This use of theory, as a guiding framework generated through the use of literature and data analysis, is common in much ethnographic work (see, for example, Layder 1998).

Dilemmas about knowledge

Public knowledge versus private knowledge

Generally speaking, especially within western cultures, teaching involves the idea that there is a body of knowledge, and a set of procedures and skills, that learners must or need to acquire. At the same time, there is also a common belief that knowledge is only useful and valuable when it enables people to make sense of their own experiences, and that for this to be possible it must be constructed by the learner. The relationship between these two views, when they are held together, generates a dilemma, notably about the role that learners' own beliefs and understandings should play in lessons. From an extreme, 'transmissionist' view of teaching, these have no role at all, they are to be replaced by what the teacher provides. For many constructivist science education researchers, the role of learners' prior beliefs must be recognised as obstacles to learning, and strategies used to overcome them (see for instance, Osborne 1996). For more progressivist views of teaching they can and must, at the very least, be the starting point for learning (e.g. Driver 1983; Harlen and Qualter 2006).

Like many teachers, Coral felt the force of this dilemma. Thus, she regards science as a body of knowledge and a way of thinking about the world, what she calls 'logical thinking' (interview data), that must be introduced to the children. Indeed, she places particular emphasis on children learning 'logical thinking', by which she means the processes and procedures that scientists employ. At the same time, she argues that children bring their own ideas to the science classroom, which are the outcome of their attempts to solve problems that they encounter in their everyday interactions with the physical world, and that she must take these into account. She assumes that children's everyday ideas about the world are likely to be different from those of scientists, and that their *ways* of thinking will differ from the 'logical thinking' that scientists employ, at least to some degree. In her science lessons, she is particularly interested in eliciting and identifying the children's everyday ideas and uses a variety of methods to do this. She believes that her role is to help children understand their own experiences in a scientific way, to become 'really scientifically literate' (interview data). She wants children to be able to look at their everyday experiences and say: 'I know what is happening there. I understand it' (interview). Referring to a session on gravity and air resistance, she explains that although children had their own ideas about falling objects, they had not 'thought about it like Galileo or Newton'. As a result, encouraging the children to think 'logically through it, they were surprised when they found that things fall at the same speed' (interview). In her teaching she seeks to elicit children's ideas about a scientific topic, so as to take account of these in her teaching, and to assess their understanding in relation to the scientific knowledge that she wants them to learn.

However, doing this is not a simple matter. On one hand, Coral is aware that not all of the children's ideas are 'firm or deeply held beliefs' (data from Coral's diary), and therefore not all should be given great weight. On the other hand, she argues that not only is it difficult to access children's thinking but that she often finds that she cannot explore further even those of the children's ideas which seem to be a product of their 'real thinking'. Children think 'too much, too fast' she wrote in her diary, and, therefore, she cannot take account of many of them. For instance, during the session on gravity and air resistance, one child asked a question about the direction of air resistance, which was not discussed further; even though later, she noted that 'it could have led to an interesting discussion about directional forces, and air resistance opposing movement rather than gravity' (diary). Not being able to explore all of the children's ideas is experienced as a problem for Coral. As she points out, on the one hand she professes interest in the children's ideas yet on the other hand she 'ignores almost all of them' (diary). This tension is very central to Coral's experience of teaching. Arellano et al. (2001) argue that such tensions are associated with issues about 'authority of knowledge' with questions about who determines what ideas are validated in classroom science learning.

Hierarchical structure of knowledge versus no hierarchy

This dilemma refers to assumptions about the extent to which some knowledge is inherently more difficult than other knowledge and cannot be learned until the latter has already been mastered. Coral disagrees with the idea that knowledge of a science area can be broken down into small bits and taught according to their degree of difficulty:

> I don't believe that you can put things in order, that you have to learn this first and this second and this last. Learning science does not come in nice neat little packages that can be taught in a nice neat way and fit together afterwards. (email communication)

Nevertheless, she points out that certain areas of science (e.g. forces) include difficult concepts and explanations, which perhaps cannot be mastered by primary school children:

> Well some areas are quite complicated. The teaching of forces is very difficult. The ideas are very sophisticated and maybe are not appropriate for ten year olds. It's very hard to teach forces and it's very difficult for children to understand. (email communication)

She tries to deal with this tension by structuring her lessons partly according to what she thinks is the order that children will find easiest to follow, basing this on her formative assessment of the children's current understanding. As she explains, she decided to start her teaching of the topic of forces with a session on falling objects because it is an everyday phenomenon that all children have experienced, and about which they are likely to have developed ideas. When the children had shown evidence in their understanding that they 'could expect things to fall at the same speed' (diary), Coral moved on to discuss the concept of air resistance, hoping that the children would find this phenomenon interesting to explore, since it is closely associated to gravity and the speed by which different objects fall. Nevertheless this strategy does not necessarily resolve all of Coral's concerns about accessing and incorporating in her teaching children's thinking:

> I now assumed that the children knew that they could expect things to fall at the same rate, and that they would, therefore, find air resistance an interesting phenomenon. On reflection this is a ridiculous assumption. (diary)

This illustrates the fact that while teachers 'resolve' the dilemmas they face in practical ways, there is an important sense, in which they remain unresolved, to be faced recurrently during the course of classroom teaching (see Osborne 1997).

Knowledge as content versus knowledge as process

This dilemma reflects well-known debates in the science education literature about the extent to which science should be taught as a body of knowledge about the physical world or as a process of thinking through which scientific ideas are developed, elaborated, tested, revised and so on (see Gott and Mashiter 1995).

Coral describes her science lessons as 'genuine processes of exploration' (interview), during which she helps the children in her class to ask their own scientific questions about a particular experience, and to develop their procedural and conceptual scientific understanding, so as to be able to provide adequate answers to those questions. She stresses the interrelated nature of the concepts, procedures and processes of science, and argues that while she is very interested in the content of science she does not believe that the teaching of content should 'necessarily take place at the expense of process' (interview). These views are also close to her beliefs that science teaching can *empower* children; it offers them opportunities to draw on skills, experiences and knowledge that they already possess and use in their everyday lives, and to perform well at school. From her perspective, science teaching contrasts with the teaching of maths and language, in this respect, where children have already-formulated views about their abilities:

> Science gives all the children the chance to shine at it, to discover abilities they didn't think they've got and to be the one with the good idea. In maths lessons they are never the ones with the good ideas because they learned over the years, rightly or wrongly, that they are not good at that so they keep quiet. But they don't do this in science, it is a new thing, it is a new topic. It is new to them so they all want to have a go. (interview)

Implicit in Coral's orientation is the belief that 'logical thinking' is comparable in all disciplines, and therefore that it is very important for children to master it. As she puts it: 'testing things and fair testing and logical thinking like that, I think it helps them in absolutely everything else, in the way they approach every subject' (interview). This makes an emphasis on process at least as important for her as that the children learn the appropriate scientific knowledge.

For this reason, Coral is very critical of the English National curriculum and the tests that are used to assess children's science learning (SATs). She sees these tests as simply testing children's knowledge of scientific vocabulary, and their ability to answer particular questions expressed in a decontextualised way. By contrast, she argues that the evaluation of children's learning should be more concerned with understanding children's thinking: their ability to use a scientific term is less important than their ability to explain a particular phenomenon using their own language. At the same time, she has to prepare children for the tests and she finds that her teaching changes as a result of this:

> My teaching changes all the time and it does become more content oriented. I do more that the children cannot understand really because I've been judged. I've been trying to hammer certain things: 'You need to know about this'. It's awful! (interview)

The tension here, therefore, is between Coral's view that scientific learning is achieved through children's development of 'logical thinking', and external demands placed upon her to organise her teaching in ways that ensure the effective preparation of children for formal testing. At the same time, there are ways in which this simply reinforces a dilemma that she experiences within herself.

Dilemmas concerning the nature of children's science learning

Learning by doing versus learning by hearing about

This dilemma captures the tension between Coral's belief that learning occurs when children are engaged in practical activity and the assumption that learners are capable of acquiring new knowledge by listening to their teacher or to what other children say. For Coral, children's participation in practical activities plays a crucial role in their learning because it enables them to reflect on the viability of their existing understandings, and encourages them to begin to reconsider and develop these understandings towards scientific ones. Implicit in this view is Coral's belief that learning occurs when children are intrinsically motivated, that is when they become 'genuinely curious' (email) about aspects of the topic under study. And this is achieved when they are offered opportunities to pursue their own interests. Discussing children's practical work, Coral points out that such explorations are similar to young children's play:

> [Involvement in practical activities] is not superficial messing about, it is real play like small children do when they are completely involved in what they are doing. (email communication)

In this respect, Coral's views of how children learn have similarities with Piagetian staged views of the learner, according to which the learner is an active constructor of meaning, responsible for organising, structuring and restructuring experience via existing conceptual structures and for determining the value of a conceptual structure by judging how well it fits with experience (see Driver 1984, 1989). At the same time, Coral points out that she does not necessarily expect the children to discover scientific concepts through their engagement in practical activities. For her, participation in such activities simply provides the initial basis upon which scientific knowledge can be built; it is a jumping off point that teachers can employ to facilitate learning. In this respect, her views about the learning process are closer to Vygotskian accounts, which argue that learning can only take place within learners' zones of proximal development (see for example, Harlen and Qualter 2006; Rogoff and Gardner 1984). For example, she argues that her role is to 'lead the children along as far as they can towards developing their own understanding' (interview). Coral explains that the development of children's scientific understanding is achieved during classroom discussions where she tries to identify those of the children's ideas that are closer to the desired scientific one, and to represent these ideas to the children using appropriate scientific terms:

> And then you can [after children have been involved in practical activities], I hope, build on that rather more formally and draw what I feel I can draw out of it, so they've got some real real science out of it. (interview)

Nevertheless, Coral points out that children's construction of their picture of science is often not as continuous or smooth as she would like it to be: 'It doesn't build up a nice picture, a nice neat jigsaw that's in my mind' (interview).

Quite often children forget or misunderstand what was discussed until they 'gradually build one thing on another and things come together for them' (interview). At issue here is the extent to which Coral's use of a range of approaches to science learning contributes to the development of children's scientific understanding. The picture becomes even more complex when she suggests that children's science learning can happen independently from the structure of the session.

It occurs when 'the moment is right and the interest is there' (interview). We are back here to an echo of Piaget.

Learning as social versus learning as individual

This dilemma captures alternative orientations towards how science learning best proceeds. On the one hand, Coral sees learning as an individual activity, a conceptual process during which children reflect on and begin to reconsider their existing understanding: 'The more working out for themselves they do the more conceptual development takes place in a way' (interview).

Within this context, social interaction is external to children's development of scientific understanding. For instance, Coral argues that children are capable of developing each other's scientific thinking when they are working in pairs because 'they talk about what they are doing and they get more out of it' (email). However, she does not believe that encouraging children to work in groups bigger than two necessarily promotes science learning. This is because she believes that children at this age (9 and 10 years old) are still 'socially egocentric' (interview); that is, they do not seem to have skills of collaboration. Therefore, when they are working together in a large group, they find it difficult to listen to and discuss each other's ideas. To support her view, Coral gives an example from classroom discussions. She says that children quite often offer responses to a question that she asked earlier in the session, despite the fact that the theme of the current discussion has changed:

> I don't particularly like big groups. Threes I tolerate. But bigger than that, children are not involved. I think children are pretty egocentric at this age, nine or ten. They will come with an answer to the question or to the idea that occurred to them five minutes before and they will say *that* regardless of what has gone on since. And I think they are like that with each other as well. They are very much into their own track. (interview)

However, while Coral does not promote group discussions, she does very much encourage whole-class discussions, in which she organises the discussion and regulates the exchanges of ideas among the children and between the children and herself. She believes that in this way she is able to help children develop one another's ideas. She describes the outcome of such discussions in terms of 'shared thinking' (interview), and argues that this often occurs implicitly:

> Most of the learning that takes place in my science teaching is the result of shared thinking. Children do develop other children's ideas, without even realising it sometimes. Sometimes, I say 'this was my idea' and they will say 'no that was my idea' and it was, but the two things were so linked. (interview)

Furthermore, Coral also believes that learning science can occur indirectly, when meanings are not explicitly negotiated between the teacher and the children. She argues, for example, that it is possible for the children to build up their 'picture of science' (interview) when experiencing situations outside the school classroom.

Dilemmas concerning Coral's role in her teaching

Teacher versus child control

This dilemma captures the pull between, on the one hand, the teacher exercising control over what children are to learn, and how, during a science lesson; and, on the other, allowing them to exercise control over their own learning. For Coral, this amounts to a tension between allowing children to pursue their own interests during a practical activity and controlling what happens so as to achieve her own learning objectives: 'I think that my learning objectives for them are important, and I suppose I feel that *their* learning objectives are secondary' (interview).

To some extent this is encouraged by her belief that although many children are able to develop their conceptual understanding with little guidance from her, many others are unable to do so, which means that she needs to exercise more control over their learning:

> There are some children whom I feel I can't trust to have the development of a conceptual framework as a primary aim. The most straightforward way of making sure that these children use their time effectively is to control what they are doing by giving them instructions from which they cannot vary unless I agree. (diary)

In her teaching, she tries to deal with this tension by offering all the children a certain degree of autonomy to modify a specific task by pursuing their own ideas. And, interestingly, this is very much what Berlak and Berlak found in their own investigations of 'progressive' primary classrooms: rather than the children being totally in control of what happened, they exercised autonomy within a clear framework laid down by the teacher. As Coral comments, a situation where all the children in the classroom perform the same activity in exactly the same way would be 'pointless' (diary), but so too would be allowing them completely free range. Instead, she tries to work in such a way as to encourage the children to pursue those of their ideas that seem closer to her own objectives. This enables her to maintain control over what is learned while at the same time giving the children some scope for exploring their own experiences and thoughts. For example, during the first session on gravity and air resistance, children were given two templates for spinners and asked to investigate how to make paper fall as slowly as possible. One child found out that a piece of paper the same size with no cutting or folding fell more slowly than either spinner. Coral regarded the child's finding as 'interesting' (diary) and decided to ask her to repeat the test in front of the class, thinking that it could lead to a discussion about rate of fall and surface area, which was one of her learning objectives for that session.

Moreover, on some occasions, children may begin to explain adequately a particular phenomenon (e.g. the rate of fall of parachutes), drawing on different concepts from the ones included in Coral's learning objectives. To some degree she encourages this, especially in relation to 'the more able and creative children' in her classroom (diary). However, the decision about whether to explore further such ideas seems to be based on what she thinks is an appropriate line of development for the majority of the children in the class. For example, as mentioned earlier, during the session on gravity and air resistance, Coral wanted the children to begin to develop their understanding of the relation between 'air resistance, surface area, density and rate of fall' (diary). To achieve this aim, she asked the children to design their own parachutes according to their own criteria. Two children who were working together seemed to be thinking of the effect of shape on the falling of parachutes and appeared to favour a round shape for their own design. Coral did not know about their theory until the following session but she argued that even if she had known of it she would not have used it, because she did not believe shape was a suitable idea to take into account for the majority of the children in her class:

> Shape was not a fruitful variable on my agenda – I wanted them to think about surface area and density. Also, round is a difficult shape – we don't know how to calculate the area of a circle and we are not very good at using compasses. Using shape as an independent variable would have been very difficult mathematically in terms of keeping the area constant. It would have led us into aerodynamics rather than surface area and density. (diary)

As mentioned earlier, in her teaching, Coral tries to balance her beliefs about how children learn science with her concern to help all the children develop appropriate scientific understanding.

Conclusion

In this paper, I set out to explore how useful the language of dilemmas developed by Berlak and Berlak is in describing the perspective and practice of an expert science practitioner. I have presented a range of dilemmas that capture tensions over how scientific knowledge should be presented to children, how children learn science and how science teaching should proceed. Some of these dilemmas are generated by the relationship between the teacher's own ideals and the external requirements placed upon her, but others reflect important tensions within her own values and commitments. Dilemmas provide us with the lenses for inquiring into how an expert primary science practitioner negotiates among such competing pressures.

It is common for researchers to view the behaviour of teachers as resulting from commitment to some single, coherent general orientation or ideology, for instance of a traditional or more progressive kind, and also to see social change as requiring that teachers adopt a new perspective, for example based on constructionist ideas about knowledge and learning. However, my analysis here indicates that we must view pedagogic practice as more dynamic in character. This is partly because teaching is *pragmatic* in character, in the sense that it involves making judgments *recurrently* about what is happening and what is needed in order to further children's learning (see Olson 1992; Schwab 1969). Moreover, teachers typically deal with whole classes of children simultaneously, and this heightens the need for contextual judgments in which conflicting considerations are balanced against one another.

In my original analysis, I used the concept of *bricolage* to explain how the expert practitioner in my study furthered children's learning by inventing solutions to problems, often on the spot, drawing on cognitive and other resources available to her (see Traianou 2006). The teacher as a bricoleur focuses particularly on how teachers *act* in order to resolve problems in their practice. Berlak and Berlak's language of dilemmas provides a different perspective for exploring the dynamic nature of teaching. It stresses the competing pressures under which teachers operate, as reflected in different aspects of their teaching. By emphasising the complexity of teaching, both dilemmas and *bricolage* should be seen as complementary tools for understanding not only how teachers *act* but also how they *think* about and *reflect* on their teaching and the possibilities for change it.

The findings of ethnographic research of the kind that discussed in this paper can help us to re-envision the education of science teachers. For example, in preparing teachers for work in primary science it is important to engage them in inquiries about their own views about the learning and teaching of science, as well as to expose them to a range of learning theories, to research on the learning and teaching of science and to a variety of teaching approaches. Such explication and exposure will help them to begin to develop a clearer sense of the kind of teacher they want to become and to recognise that tensions and dilemmas play a crucial part in shaping their identities. A dilemmatic approach to the education of teachers may also be useful for tutors working with clusters of student teachers, or with practising teachers in their classrooms, facilitating discussion of the ways in which they employ scientific knowledge and pedagogical resources to deal with the contingencies of practice.

The main emphasis in such discussions must be to encourage teachers to recognise the distinctive character of the dilemmas and tensions that they experience in their teaching and to develop the skills of the bricoleur and the reflective practitioner: the ability to be reflective in and about their teaching, and to look for new or different ways of improving their science understanding and their practice. This approach to science education is particularly important in the current climate of government-led interventions, in the UK and elsewhere that seek to pre-specify what teachers ought to know and how they ought to teach science. Indeed, it is essential.

References

Adler, J. 2001. *Teaching mathematics in multilingual classrooms.* London: Kluwer Academic Publishers.

Arellano, L.E., L.T. Barcenal, P.P. Bilbao, A.M. Catellano, S. Nichols, and D.J. Tippins. 2001. Case-based pedagogy as a context for collaborative inquiry in the Philippines. *Journal of Research in Science Teaching* 38, no. 5: 502–28.

Berlak, H., and A. Berlak. 1976. Toward a political and social psychological theory of schooling: An analysis and interpretation of English informal primary schools. *Interchange* 6, no. 3: 14–5.

Berlak, H., and A. Berlak. 1981. *Dilemmas of schooling: Teaching and social change.* London and New York: Methuen.

Billig, M., S. Condor, D. Edwards, M. Gane, and D. Middleton. 1988. *Ideological dilemmas: A social psychology of everyday thinking.* London: Sage.

Chaiklin, S., and J. Lave, eds. 1993. *Understanding practice.* Cambridge: Cambridge University Press.

Denzin, N.K., and Y.S. Lincoln. 1994. Introduction: Entering the field of qualitative research. In *Handbook of qualitative research*, ed. K. Denzin and Y.S. Lincoln, 1–17. Thousand Oaks: Sage.

Driver, R. 1983. *The pupil as scientist?* Buckingham: Open University Press.

Driver, R. 1984. A review of research into children's thinking and learning in science. In *Learning, doing and understanding in science: The proceedings of a conference* SSCR, ed. B. Bell, W. Watts, and K. Ellington, 8–21. London: Woolley Hall.

Driver, R. 1989. Students' conceptions and the learning of science. *International Journal of Science Education* 11: 481–90.

Edwards, D., and N. Mercer. 1987. *Common knowledge.* London: Routledge.

Gott, R., and J. Mashiter. 1995. Practical work in science – A task-based approach? In *Subject learning in the primary curriculum*, ed. P. Murphy and M. Selinger, 184–96. R. Milton Keynes, UK: Open University Press.

Hammersley, M. 1977. *Teacher perspectives.* Milton Keynes, UK: The Open University Press.

Hammond, L., and C. Brandt. 2004. Science and cultural processes: Defining an anthropological approach to science education. *Studies in Science Education* 40: 1–47.

Harlen, W. 1999. *Effective teaching of science: A review of research.* Edinburgh: The Scottish Council for Research in Education.

Harlen, W., and A. Qualter. 2006. *The teaching of science in primary schools*, 5th ed. London: David Fulton Publishers.

Jaworki, B. 1994. *Investigating mathematics teaching: A constructivist enquiry.* London: Falmer.

Lampert, M. 1985. How do teachers manage to teach? *Harvard Education Review* 55, no. 2: 178–94.

Lave, J., and E. Wenger. 1991. *Situated learning: Legitimate peripheral participation.* Cambridge: Cambridge University Press.

Layder, D. 1998. *Sociological practice: Linking theory and research.* London: Sage.

Lévi-Strauss, C. 1966. *The savage mind.* London: Weidenfeld & Nicolson.

McGinnis, J.R. 2006. Cultural considerations. In *Elementary science teacher education*, ed. K. Appleton, 275–98. London: Lawrence Erlbaum.

McGinnis, J.R., C. Parker, and A. Graeber. 2004. A cultural perspective on the induction of five reform-minded new specialist teachers of mathematics and science. *Journal of Research in Science Teaching* 41, no. 7: 720–47.

Nelson, C., P.A. Treichler, and L. Grossberg. 1992. Introduction. In *Cultural studies*, ed. L. Grossberg, C. Nelson, and P.A. Treichler, 1–16. New York: Routledge.

Newman, E. 1997. Primary teachers' dilemmas in initial teacher training. *Teacher Development* 1, no. 2: 253–68.

Olson, J. 1992. *Understanding teaching.* Milton Keynes, UK: Open University Press.

Osborne, J. 1996. Beyond constructivism. *Science Education* 80, no. 1: 53–82.

Osborne, D.M. 1997. Balancing individual and the group: A dilemma for the constructivist teacher. *Journal of Curriculum Studies* 29, no. 2: 183–96.

Osborne, J., and S. Simon. 1996. Primary science: Past and future directions. *Studies in Science Education* 26: 99–147.

Rogoff, B., and W. Gardner. 1984. Adult guidance of cognitive development. In *Everyday cognition: Development in social context*, ed. B. Rogoff and J. Lave, 95–116. Cambridge Mass. Harvard: Harvard University Press.

Roth, W.-M. 1999. Authentic school science: Intellectual traditions. In *Learning and knowledge*, ed. R. McCormick and C. Paechter, 95–116. London: Paul Chapman Publishing.

Roth, W.-M. 2000. Autobiography and science education: An Introduction. *Research in Science Education* 30, no. 1: 1–12.

Roth, W.-M. 2005. *Auto/biography and auto/ethnography: Praxis of research method.* Rotterdam: Sense publishers.

Schwab, J.J. 1969. The practical: A language for curriculum. *School Review* 78, no. 1: 1–24.

Summers, M., C. Kruger, and J. Mant. 1997. *Teaching electricity effectively: A research-based guide for primary science.* Primary School Teachers and Science (PSTS) Project. Hatfield, UK: The Association for Science Education.

Traianou, A. 2006. *Understanding teacher expertise in primary science: A sociocultural approach.* Rotterdam: Sense Publishers.

Traianou, A. 2007. Ethnography and the perils of the single case: An example from the sociocultural analysis of primary science expertise. *Ethnography and Education* 2, no. 2: 209–20.

Wertsch, L.S. 1985. *Vygotsky and the social formation of mind.* Cambridge, MA: Harvard University Press.

Feeding pigs and looking for *güembé*: the local production of knowledge about the natural world of peasant and indigenous children in San Ignacio

Ana Padawer

Facultad de Filosofía y Letras, Instituto de Ciencias Antropológicas, Consejo Nacional de Investigaciones Científicas y Tecnológicas (CONICET), Universidad de Buenos Aires, Buenos Aires, Argentina

Ever since the first ethnographic studies, the question about how people learn in different contexts was an essential element of research. Studies in anthropology of education have increasingly focused on schooling. Nowadays, however, the traditional topic of learning in out-of-school contexts recovers its relevance because the development of technologies, globalisation and inequality create new contexts for knowledge. Peasant and indigenous children's productive activities in countryside are frequently defined as children's labour, tending to emphasise the obstacles to attending school and ignoring the production of knowledge about natural world. In this article, I will use concept of legitimate peripheral participation to identify situated learning outside school walls. This category allows me to describe the understanding of natural and social world that peasant and Mbyà-Guaraní children in rural areas of San Ignacio (Misiones, Argentina) achieve in their everyday learning experiences. Information about plants, insects, animals, birds, their customs, food preferences, reproduction methods and environment are learned by observing and doing, while science lessons usually ignore such familiarity.

1. Introduction

Peasant and indigenous children habitually carry out productive activities in the forest and countryside as part of family reproduction. However, both scholars and common sense do not usually recognise these activities as implying knowledge. In this article, I will use the concepts of legitimate peripheral participation (Lave and Wenger 1991), communities of practice (Wenger 2001) and ethno-science (Ellen 2004) to identify situated learning outside school walls. These categories allow me to describe the knowledge of the natural and social world that peasant and *Mbyà-Guaraní* children from rural areas of San Ignacio (Misiones, Argentina) attain in their everyday learning experiences. Information about plants (their characteristics, properties and uses), and animals (food preferences, reproduction and environment) are learned by observing and doing, while science lessons usually tend to ignore such familiarity.

However, I will also explain how this local knowledge is decreasing in the learning experiences of new generations, as productive activities decline. In the last decades, the concentration of land ownership has increased in this area; although this situation has roots in an unequal distribution of land, which began in the last decades of the nineteenth century, forest exploitation and a recent crisis in the yerba mate industry have had dramatic consequences on families' ways of living. On one hand, this situation has restricted the mobility of *Mbyà-Guaraní* people even more, due to the fact that land is now highly valued for forest exploitation, and communities usually do not have legal documents of ownership. On the other, many peasant families have been forced to abandon territories that they have used for generations, or have also been compelled to sell their properties – if they have them. Urban migration has been dramatic for peasants: unemployment is high in San Ignacio, a city of 10,000 inhabitants, whose main income is generated by tourism.

My fieldwork shows that among some peasant families (tenant farmers and small landowners, which have been able to remain in the countryside), children can still learn to take care of farm animals and to carry out agricultural tasks, although such activities tend to decrease from one generation to another. Because some farmers and most *Mbyà* communities do not have legal ownership of the land where they live, they often have to move and their horticulture is highly affected. Among indigenous families, limited access to the native forest has damaged the transmission of expertise on to new generations, especially knowledge regarding animals and plants traditionally recognised for nutrition, health and craft. Nevertheless, *Mbyà-Guaraní* people persist in their forest incursions and political struggle for territories, ensuring their subsistence with some customary practices of crop growing, hunting and gathering, activities that they supplement with State assistance.

This article is based on fieldwork conducted since 2008 in a zone defined by the highest concentration of M*byá* communities in San Ignacio Department (Centro de Trabalho Indigenista [CTI] 2008), including 9 of 16 small villages with a population between 20 and 100 people. These villages are located near the city (no more than 40 km away), so they are in a rural space undergoing considerable transformations because of economic activities.

Despite having information regarding eight of nine villages (Andresito, San Ignacio Miní, Katupyry, Kokuere'i, Pindoity, Ñu Porá, El Tacuaral and Ivy Poty), this article focuses on the first of them: Andresito. This community is relevant in educational terms because its children are attending a rural school, which peasant children also attend (most of *mbya-guarani* communities have their own schools in their villages). As a result of this process, half of the students of this school are *mbya* and the rest identify themselves as *colonos* (people who live in the frontier areas). The school is functioning with three teachers for basic education (multigrades), one for kindergarten, and two indigenous teacher's assistants; they have almost 90 students.

As with *mbya guaraní* villages, for the *colonos* I have defined a zone to be studied near the city head of Department of San Ignacio. There are eight major towns in this department, most of them founded in the 1950s along the traces of a national route, but two cities are older: San Ignacio was founded in 1693, and Corpus in 1882 (Gobierno de Misiones 2008). Almost 60% of the population of San Ignacio is urban, but this study is located in the rural area, at *las colonias* (place where the *colonos* live). The peasants mentioned in this article live in *las colonias* nearest San Ignacio town, known as Aparicio Cue e Invernada.

Ethnographic fieldwork included participant observation and interviews; some classes of children between ages 10 and 12 were observed, but also I used some experimental techniques in order to work with children and their everyday activities: drawings, photographs and video, free listing. All families of the aforementioned school were visited in order to interview adult members and reconstruct economic activities. *Mbya-guarani* leaders of the community of *Andresito*, indigenous teachers and school staff, local authorities and state administrators were interviewed as well.

2. San Ignacio: productive activities in the forest and countryside for family reproduction

San Ignacio is 1 of 17 departments of the province of Misiones (in the Northeast of Argentina). It borders Paraguay and Brazil, and is nowadays the region with the largest density of *Mbyà-Guaraní* people in the country: about 4083 members of this indigenous community, according to Complementary Survey of Indigenous People conducted in 2004–2005 by the National Institute of Statistics and Census (Instituto Nacional de Estadística y Censos [INDEC] 2010).

The Guaraní occupation of the current territory of Misiones is very early: historical data suggests that at the beginning of the sixteenth century this ethnic group was going through a geographical and demographic expansion, with a population of over two million.[1] Their population spread through the delta of the Río de la Plata in the current territories of Argentina and Uruguay, the shores of Santa Catarina and Paraná, Curitiba and Mato Grosso do Sul in Brazil, and largely in the current territory of Paraguay. From the first contacts with Europeans towards 1513, the population diminished dramatically because of the introduction of infectious diseases, regional wars and slavery (Noelli 2004, 17).

The Guaraní, considered the cultural, linguistic and demographic foundation of the region's contemporary indigenous population, include four subgroups traditionally named *Mbyà, Pai-Taviterá, Avá-Chiripá y Aché-Guayakí.*[2] The current *Mbyà* are thought to descend from the populations that remained independent from the Jesuit colonial experiment between the sixteenth and eighteenth centuries, by building small communities in the forest that involved the construction of a social identity defined by the confrontation and contrast with the Spaniard conquerors first, and their *mestizo* descendants later (Bartolomé 2004).

Subsequently, the establishment of the *Mbyà* in current Argentine territory took place in interrelation with the expansion of the national society's frontiers during the nineteenth century and the first decades of the twentieth century, causing a progressive dispersion and reduction of the villages throughout time. This dispersion however, has not prevented the recognition of a political adherence to a distant authority in some cases (Gorosito 2005). This political dimension is closely related to the current economic constraints on the reproduction of families and thus, to the intergenerational transmission of resources. As we will see below, a dynamic process of creation of small *Mbyà* villages is currently taking place in the area of San Ignacio. These villages stem from other larger ones, and frequently emerge in smaller extensions of land, generating new political leaderships.

The history of the peasant settlement in the area is connected to the aforementioned process of frontier expansion: during the last decades of the nineteenth century, a colonisation process organised mainly by the State took place in the area of Misiones, on the lands leftover after a massive sale to 40 large buyers. However, the occupation and exploitation process that began as an agricultural frontier was not completed only through the colonisation organised by the State; it was also a spontaneous process, where the settlement of land took place in association to companies with forestry operations, which allowed the land to be used after the extraction of wood was completed (Ricotto and Almeida 2002).

Until 1930, farmers only produced yerba mate; in turn they incorporated tung –a tree used for extracting oil to industrial purposes, tobacco and tea, which were grown at the same time forest exploitation was practiced, first of native species and later of exotic ones. In this way, an agrarian society developed in Misiones during the twentieth century, formed by peasants and squatters (household agricultural producers with 1–10 hectares [ha], of *criollo* origin or Brazilian or Paraguayan immigrants);[3] farmers (with 25–50 ha, mostly of Northern and Eastern European origin);[4] stockbreeders (with land between 100 and 1000 ha) and large extractivist landowners (Bartolomé 2000; Jaume et al. 1989; Otero 2008; Reboratti 1979).

If we use this reconstruction to analyse the information provided by the National Agricultural Census from 2002, we observe that, nowadays, in the department of San Ignacio almost 30% of the productive area corresponds to typical squatter, peasant and farmer properties, more than 45% to realties defined as stockbreeding ranches and over 25% to large estates (Gobierno de Misiones 2008, 340–41).

The current situation is based on the structure of the Misiones agrarian society consolidated during the twentieth century, and it reflects the changes that took place in the last decades. Different studies show that these changes had to do with the fracture of the institutional political structure, and with the capitalist accumulation process of the Argentine society installed in 1976. This process led to an agricultural crisis and a decline of the peasant economy – along with changes in the concentration of land ownership – that accelerated the process of depeasantisation (Cragnolino 2006; Manzanal and Rofman 1989). In the last years, along with the repositioning of agricultural products in the world market, the predominant economic model shifted towards agro-industrial production, affecting small producers and indigenous populations differentially in their access to resources, work and survival (Bidaseca and Mariotti 2001; Sili 2005).

The agro-industrial complex had been expanding significantly in the province of Misiones since the 1980s, especially in the tobacco industry. The tobacco chain involved farmers and squatters that currently grow this crop, while maintaining certain productive diversification – stockbreeding, forestation, production of yerba mate and horticulture. Despite the growth of the agro-industrial area, these producers have become impoverished due to the process of concentration of manufactures and the commercialisation of crops by stockpilers and millers, both in the case of tobacco and *yerba mate*. In addition, in the last 20 years conflicts with the owners of native forests have multiplied. These owners had once facilitated the use of their land to farmers and squatters; however, with the development of a timber industry of exotic species led by concentrated capital, they claim the highly valued land (Schiavoni 2008).

These processes that took place in the last 30 years also affected *Mbyà* communities: while villages could remain in relatively less exploited areas of the forest, they could articulate their trade relations with the regional society and maintain relative control of their own organisational processes and mobility.

Nowadays, the changes in the social and economic structure of the province, as well as the relations with the State in the last two decades, have lead to a significant multiplication of loosely established autonomous residential units, with weak political organisation alliances and an increased dependence on the regional economic structure (Gorosito 2005).

Depending on the proximity to non-indigenous populations, *Mbyà* communities located in the territory of Misiones currently carry out diverse subsistence activities that include the hunting of small animals, fishing, seasonal gathering, vegetable garden production and rearing of fowl and pigs. In some communities members also participate in salaried work in agriculture and in activities linked to tourism – such as selling crafts, tour guides through the woodland and visits to villages (Cebolla Badie 2005).

Despite the weakening of political organisation caused by the aforementioned dispersion, the last years have witnessed a gradual acknowledgement of indigenous communities. They claim and exercise citizen rights as a particular political community, thus producing shifts in the economic, political and formative activities of younger *Mbyà* generations. Among these changes, Wilde (2007) points out how environmentalism and indigenism as political articulations present tensions in Misiones: occasionally, an incongruity is presented between the 'ecologically noble savage' – a stereotype defended in environmental discourse – and the inevitable commercial insertion of many indigenous groups. This incongruity is linked to the misguided assumption that communities are a homogeneous whole, represented by leaders of unquestioned legitimacy.

In addition to land claims, one of the main demands of indigenous populations refers to forest exploitation control. Following the extractive process of native species – which is currently regulated by the State – land is usually treated with herbicides, and pine plantation follows. These processes affect indigenous dwellers, farmers and squatters by polluting watercourses and because the progressive deforestation restricts the hunting and gathering activities of the *Mbyà*. Moreover, large-scale productive activities use scarce workforce, so there are limited opportunities for salaried work as a resource for survival.

3. Legitimate peripheral participation and situated learning outside school

Children's participation in household productive activities can be understood as a formative experience if educational processes are conceived as contextually situated, in a way that learning is produced through communities of practice and legitimate peripheral participation (Lave and Wenger 1991; Wenger 2001).

The concept of legitimate peripheral participation is closely related to the concept of guided participation (Rogoff et al. 1993, 6). The latter, drawing from Vygotsky's developments, has lead to the reformulation of the study of children's knowledge in different sociocultural contexts. According to this perspective, children's understanding grows by means of a creative process through which they transform what they know and their own world, while progressively participating in their communities' activities.

Meanwhile, legitimate peripheral participation refers more directly to knowledge by doing, based on a reformulation of the term apprenticeship. This concept stems from the debate on the nature of learning in the late 1980s, and assumes that learning is always situated: this implies not only that learning takes place in time and place, with other people, or depending on the context in which it is produced; it also emphasises its character as a situated activity. Therefore, the concept of legitimate peripheral participation describes the involvement in social practices that are themselves constituted by learning processes, and not vice versa (Lave and Wenger 1991, 33).

According to the above, even so called general knowledge – which is usually presented as opposed to situated knowledge – emerges from abstract representations that acquire meaning in a context, and that are themselves achieved in specific circumstances. Hence, the notion of community of practice allows the discussion of dichotomies between action and knowledge: the process of participating in a practice always involves the whole person, who is acting and knowing simultaneously. Manual activity is not unreflexive and mental activity is not incorporeal; all social subjects are understood as carrying theories and ways of understanding the world. Communities of practice are the places where these are developed, negotiated and shared (Wenger 2001, 72–3).

In addition, the idea of communities of practice, where participants can occupy centre and periphery positions, also implies that these processes involve power and hegemony relations: full participation entails a strong dominion of collective knowledge or practices that must have accessible degrees of attainment for novices. Nonetheless, the peripheral character refers to a progressive access to sources of understanding through growing involvement (Lave and Wenger 1991, 37).

It is not only about the ability of engaging in new activities or mastering new information, but also about establishing new relationships enabled by that dominion. Through these relationships the individual takes place in the production and reproduction of the structures of the communities of practice in which he participates. Like Rogoff et al. (1993), Lave and Wenger (1991) draw these ideas on learning in its individual and structural dimensions, from studies inspired by Vygotsky's notion of zone of proximal development. These concepts enable the revision of common-sense ideas that understand educational processes as non conflictive transmission and internalisation of knowledge; in this sense, it is more pertinent to understand learning as an appropriation process (Rockwell 1995, 56–7), in that it refers to the conflictive nature of social practices, in a way that the relationships between apprentices and those who are experienced are part of social transformation processes that take place on an everyday level.

With this conceptual framework, Lave and Wenger's notion of apprenticeship differs from the classic approaches to the concept, where it was understood as an activity linked to craftsmanship on an individual basis or in small groups, the use of simple tools and tacit knowledge, the division of labour based on individual adaptation and the prevalence of traditional protection codes. For Lave and Wenger, apprenticeship assumes diverse historical forms, cultural traditions, and can be found in different modes of production; it is neither inherently egalitarian nor exploitative; the point is to analyse the form of political and social organisation under which it develops in each context, and the formative principles under which it is articulated and carried out (1991, 62–3).

Given that reflection on the relationship between knowledge and social practices, it is important to note that the opposition between scientific knowledge and other ways of knowing – defined as indigenous, local, traditional, intuitive, ordinary, technical – rests on the epistemological distinction that originally constituted anthropology as a discipline, that confers truth to science over religion and superstition. Ellen (2004, 411) states that this distinction conceals the various ways in which predictive knowledge about the material world is produced, through distinctive configurations of cognitive and technical developments that are produced through historically constituted social practices. From his point of view, it is important to distinguish not whether certain knowledge is scientific or not, but rather between more versus less skilled thinking: it emerges facing difficulties or mysteries, understanding that most of the time social subjects use broadly shared answers, that have important historical depth and are effective for daily needs (Ellen 2004, 426).

In view of the above considerations, in the daily social practices of children that will be presented in this article it is possible to analyse certain attributes that are frequently ascribed to scientific practices and which are mistakenly used to distinguish them from traditional and empirical knowledge. One attribute is suspension of beliefs, which consists of events where there is no mechanical application of existing knowledge, but where people exchange differing points of view which, in turn, enable the acceptance that the world is different in some of its features. Other attributes are connection of classifications with different concepts of causality and the use of framework – not specific theories. Finally, three attributes are interesting: (1) networks of understanding – that relate individual species to specific contexts and landscapes, not isolated species; (2) experimentation – that enables the assessment of data for a series of explanations, including a methodology to collect them; and (3) construction of narratives, from which ideas emerge by analogy or other procedures (Ellen 2004, 426–28).

Finally, it is important to consider that in practices of knowledge about how the world is constructed and can be analysed, nature is a convenient fiction (in terms of a human construction and narrative), where there is an ambiguous relationship between what can be empirically observed and what is represented. This issue, which is inscribed in the aforementioned dualisms regarding the concrete and the abstract, body and mind, has been solved in the convergence of cognitive and symbolic approaches in Anthropology, by recognising that human populations apprehend both the social and the natural worlds, one in terms of the other: this is why the metaphoric extension from one to the other is not problematic and, in contrast, it is productive in terms of understanding how the world can be thought (Ellen 2004, 430–31).

With the identification of natural and social knowledge that peasant and indigenous children attain in their daily practices we do not intend to portray them as 'little scientists': science can be defined as a material practice that combines representational conscious and embodied knowledge, is carried out by adults with sophisticated instruction and explicitly produced with those ends, it involves certain processes that are not verified in the children's daily activities, such as the lexical codification of knowledge, its textualisation, inscription and formalisation, nor its application in instruments – that allow for it to be recorded, its experimental recreation, simulation and analogy (Ellen 2004, 440–41).

In this article, I will hold that this distinction between local and scientific knowledge – as well as its continuities – is relevant in terms of the development of the formative potential of children's daily experiences in the school context, because in its routines it constituently includes many processes that are distinctive of scientific practice, especially textualisation and formalisation. Recognising the possibility of inscribing formative experiences produced on a daily basis outside school, allows them to be transformed, incorporating processes of skilled thinking that are not usually identified in the productive context but that are significantly based on it.

4. Everyday lessons: knowledge of the natural world of peasants and *Mbyà*-Guaraní children by their participation in productive activities

4.1. Living species in relationship

Sebastian is 11 years old; he is the second of four children and has three sisters, one that is older than him (who attends high school in another town) and two younger ones. Marta, age 10 and Alicia, age 7.[5] Their father owns a truck and hires labourers to chop and load wood, which is why he is out of the house Mondays through Fridays. The mother and her three children stay at the farm, of 200 ha, where they currently have a pine and chinaberry tree plantation intended to revitalise a family sawmill, as well as a small farm where they have 30 cows, a herd of pigs and hens. They plant cassava (*Manihot esculenta*), corn, pumpkin and sugar cane which they use both for their own consumption and for the animals; they also grow a small garden of mainly leaf vegetables for household consumption (lettuce, chard, onions, parsley, radish, carrots).

When they arrive at home from school and after lunch, Sebastian and Marta devote approximately an hour and a half to activities on the farm, generally feeding the pigs and a few cows that are occasionally tied because they are sick or because they are meant for the production of milk for family consumption. The animals are fed with cassava and corn, which the children extract from the plantation on a daily basis.

Although this routine does not seem to involve challenges in terms of notions regarding the natural and social world, I would like to show how Sebastian and Marta deal with knowledge that comes from situated learning in communities of practice. This knowledge allows them to carry out certain tasks autonomously, drawing from their previous peripheral participation in different tasks in order to do so. The predictive capacity regarding the material world that the children handle is based on networks of understanding attained during the course of their activities, where the relationship between what is empirically observable and what is represented is integrated through narratives. Skilled thinking develops when faced with difficulties and queries; in this process, school can play an interesting role in the lexical codification, inscription and formalisation of experiences.

When observing the children's daily routine, the first thing noticed is that they do not need to consult their mother because they know which of the available foods on the farm are good for pigs: mainly corn and cassava, and to a lesser extent pumpkin and yams. Although the children pointed out that pigs like corn the most, they have not asked themselves about this preference, nor can they state the reasons for it. However, they do create a contrast and a causal relation in reference to another animal that they take care of daily and that they feed in a different way. Sebastian resorts to a narrative to contrast the feeding of pigs and cows: he recalls that when releasing a pregnant pig from her pen in order for her to give birth more comfortably, he saw that the animal actually ate grass but fattened less. For this reason, though he feeds cows with grass, he does not do so with the pigs.

In addition to identifying the crops, the children are familiar with the appropriate procedure while feeding the pigs: they partially peel the dry corncobs and throw them to the animals, stating that it is not necessary to strip them completely because they have observed that pigs can do it themselves with their teeth. Also, given that the corn husks do not rot quickly, the animals have enough time to nose the leaves out of the pigpen, or the sows can use them in creating their nests before farrowing – we will comeback to this later. Regarding cassava, the children know that grown pigs are fed with the whole vegetable, whereas it must be cut into small pieces for the piglets, otherwise it is too difficult to swallow. Even though the children are not aware of its toxicity, they note that the piglets do not eat the skins alone: possibly this information has emerged from the attempts to profit from the waste leftover from human consumption of the vegetable, and from seeing that adult animals do eat the cassava with its skin.

Marta, the oldest of Sebastian's sisters, is in charge of chopping up the cassava for the piglets. In doing so, she trains in the use of the *machete*, an instrument used frequently on the farm, which requires skilful training due to its dangerousness. Because the child has not yet acquired the necessary skills, her older brother takes over when the machete is needed for cutting the stems of the sugar cane or for separating the stems of cassava and unburying the tubers. Sebastian occupies two crucial positions for the organisation of household work, and in relation to them he acquires specific knowledge: he is the eldest child in the house and he is a boy, therefore he is in charge of the tasks ascribed to the male role, which require greater physical strength.

The link between gender, age and agricultural tasks is naturalised, hence the widespread assumption among peasant families is that girls must learn and take care of gardening, while boys – especially older brothers – must know how to take over the farm. Where this prescription is not confirmed, it is presented more as an idealised past, with which the array of particular circumstances that a family can face are confronted: in the abovementioned case, the father's work outside of the home leads to a situation where the mother is in charge of training Sebastian – and later Marta – in the use of the machete, starting with the appropriate movements but applied in the garden, where the plants are fragile and small.

The use of the *machete* requires not only specific movement skills, but also knowledge regarding the precise places each plant must be cut in ways that allow the growth of new specimens. Thus, Sebastian can show the distance from the ground that sugar cane must be cut, so that in a month's time the plant will grow back again. Or the length the cassava must be extracted, in order to preserve the stems for drying and subsequent sowing, but at the same time keeping a long enough portion of the trunk so as to extract from the ground the root and tubers that are used for family consumption and animal feeding (see Figure 1).

Sebastian and Marta's practical knowledge of the natural world that can be drawn from the aforementioned facts is in fact deeply hidden at school. In the beginning of my fieldwork and before visiting any farm, I ask the children in school to draw and write about animals they have in their homes. Sebastian draws a pig and wrote the following:

Figure 1. Two photos of Sebastian cutting and extracting cassava.

> <u>Pigs from my house</u>
> In my house we have pigs, and we have them for killing and raising.
> We have a black sow that is pregnant and a piglet and a little sow that it is for raising and to have piglets to sell.
> Pigs eat corn and cassava and chicory.
> The sow is about to give birth next June.
> <u>Pig's Eye</u>
> Pig's eye is small and black with brown
> The head is big and furry
> Ears are big, furry and thin.
> <u>Los chanchos de mi casa.</u>
> En mi casa tenemos chanchos y los tenemos para matar y criar.
> Tenemos una chancha negra que está preñada y un cachazo y una cachacita que está para criar y que tenga cría para vender los chanchitos.
> Los chanchos comen maíz y mandioca y achicoria.
> La chancha está por tener en Junio.
> El ojo del chancho.
> <u>El ojo es chiquito y negro con marrón</u>
> La cabeza es grande y peluda
> Las orejas son grandes peludas y finites

A description followed regarding legs and the way they sleep. As can be noted, when the children are asked about farm animals in a classroom context, they tend to refer to isolated species, using some brief phrases that describe them (colour of the skin, size, basic products with which they are fed) and an unspecific tree or fence as context, even though I asked them many guiding questions (e.g., where the animals live or what they usually eat). However, when analysing the children's routine we notice – just as the most recent versions of textbooks suggest – that the knowledge they have regarding which vegetables nourish certain animals is also related to information on the reproduction of those crops, the links between animal and human nutrition, and the feeding capacities of animals in different stages of their lives. In other words, they do not have knowledge on a particular species that can be exhaustively expressed, but knowledge on the relationships between species that are relevant to immediate needs.

On the other hand, it must be noted that Sebastian's observation on grass as an inadequate food for pigs, does not emerge from a process of systematic exclusion of alternatives, but from recalling a particularly conflictive situation in the process of animal rearing: parturition. Although children's participation in this situation is extremely peripheral (the child has watched his mother help the sow in difficult farrowing situations) its formative impact is far-reaching: in addition to the information presented above – that is incidental to the animals feeding routines – the child draws knowledge regarding the precise moment to release the pigs so they can give birth more comfortably. This information emerges from the ability to observe and interconnect: the enlargement of the udder is a sign for Sebastian that the sow is about to give birth, because she is ready to suckle.

As pointed out earlier, when asked at school about the pig's nutrition, the children tend to make generalisations about animal nutrition, and they do not mention feeding cycles, or differences whether the animal is male or female. However, in Sebastian's routine we can see he includes the stems and leaves of the cassava when he is about to finish his daily feeding routine of the sow. Asked about the reasons for this difference, the child explains this is done particularly during pregnancy, because the sow crushes the stems to make a more comfortable place for giving birth. This statement is interesting because it shows how practical knowledge is more complex than the knowledge the child deploys in the classroom context. It involves both an appropriation of information regarding other farrowing experiences, and accepting that animals 'nest' when they give birth to their young as general behaviour.

4.2. The interaction of contexts

Although practical knowledge is hidden at school, children's learning experience is conducted across the two situations, so knowledge from one context has consequences upon the other, even when that interaction has not been explicitly fostered by adults. During my fieldwork, I benefited from a circumstantial event: an indigenous assistant teacher, who lives on school premises, had received a fawn as a gift after visiting another *Mbyà* community. In an activity conducted with students in my initial days at school (which consisted in making lists of animals and plants they know and to write them on the blackboard), peasant and indigenous children had recognised this animal as native to the area; however, because it is a woodland animal – and access to the forest is currently very restricted – it is unusual for children to have contact with it. Thus, the gift was an opportunity for exploring children's knowledge about forest animals.

The assistant had built two yards, one for the fawn and the other for a herd of pigs. I asked the children to observe the animals and draw them, to discuss their physical features, nutrition and habits, and to write an informal text about them. Sebastian chose to draw the fawn, but he then claimed that he could not write about these animals because his family does not usually hunt woodland animals and because of that, he does not know about their habits. I suggested that he speak to the indigenous assistant's eldest son – Andrés, of whom we will speak later on. He did so, and added an epigraph to the drawing: 'it eats *reviro* (a flour paste that is also used for human consumption) and sweet potato leaves'. The following day, I visited Sebastian's house. While he chopped cassava to feed the pigs, he found deer prints (Figure 2). He showed them to me and we covered an area of the cassava plantation where the child identified prints of different sizes – which he ascribed to animals of different ages. That same day he asked both his mother and father how they should deal with the presence of these woodland animals in the plantations. They both

Figure 2. Sebastian's drawing of a fawn and Sebastian showing fawn's footprints.

agreed that the plantation was not at risk, so it would not be necessary to release the dogs during the night.

This event triggers a reflection on the construction of knowledge about the natural and social world in productive and school contexts in different ways: first, after observing and speaking about a species at school, the child's ability to detect hints that signal the presence of these animals – that otherwise would have gone unseen – seems to have increased. The reconstruction of the presence of deer, on the other hand, was complex: he looked for the prints and analysed them on the ground – their size and depth; he observed the leaves in order to determine – according to their partial or total damage – their height and age, the number of specimens and the possible moment they had entered the plantation.

In the second place, this situation led Sebastian to inquire about the interrelation with the newly detected species: did it represent a danger for man? And in this case, did it threaten their plantation? Although the query was fully solved by the adults, Sebastian's peripheral participation in a community of agricultural practice was crucial in detecting a complex situation that the parents were unaware of: the entrance of woodland animals in the plantation. Therefore, the ability to transform the social and natural world, even for a novice, is significant: through Sebastian's discovery, the definition that his family establishes with the native woodland is re-enacted.

In the third place, and linked to the process mentioned above, the fact that *Mbyà* families catch and eat woodland animals while Sebastian's *criollo* family does not, represents for the child an insurmountable practical limit. Although eventually Sebastian could ask himself about the socially created nature of nutrition practices, the classification of animals for human consumption is working here as a mark of cultural difference, probably this situation prevents questions regarding transformation of practices. Thus, scientific knowledge, with cross cultural validity, gives intercultural education a base for particular and relevant developments in this context where traditional knowledge of peasant and indigenous people are presented as mutually exclusive, scientific knowledge in school settings can promote new questions and practices.

Finally, it is interesting to consider how knowledge generated through the identification of the animal in the productive context, emerges in a school situation, but this institution does not create a context that would allow Sebastian to move forward in the process of acquiring knowledge through inscribing and formalising his experience. Maybe recording and analysing similar situations can be a starting point to combine school and productive knowledge, challenging the children with new questions about the world, through the recovery of their experiences in both contexts.

Therefore, this is not just a reflection on processes of learning natural sciences: knowledge of the world is crossed by living conditions and daily practices of household reproduction. In the case of Sebastian, Marta and their family, we observe dilemmas regarding an extractivist production and management of the native woodland. There is no doubt that the peasant's knowledge could benefit from work in the school context and from *Mbyà* indigenous knowledge, whose complexity is rhetorically recognised, but not translated in socially relevant practices such as the school *curriculum* or local economic development policies. Finally, it is important to acknowledge the relevant role played by children in generating locally relevant and scientifically validated knowledge: through their peripheral participation in productive activities, and the inscription and formalisation of these in the school context, a space of skilled thought is enabled. This thinking can question adults in their beliefs, practices and knowledge about the world.

4.3. Observation capacities and community of practices

It is in these contrasts between formative experiences where it is relevant to analyse the contributions of Andrés, the *Mbyà* child consulted by Sebastian about the fawn's nutrition. Andrés is 13 years old and he is the eldest of José's children. As was mentioned before, José is an *Mbyà* assistant teacher that lives on school premises. The child lives with his mother, his father and his siblings: he is followed by Hernán, age 11, and Paola, age 9. In addition to their house and the two yards mentioned above – where the children observed the pigs and the fawn – José cultivates a small vegetable garden where he has planted spring onion, chard, onion and in another area, cassava and sweet potato. He has plans to begin beekeeping activities, because of training and previous practice in his native community located nearby.

In contrast to Sebastian and Marta, Andrés does not tend animals on a daily basis: this is probably linked – among other reasons – to the fact that his father is at home every day, and that their household production is significantly smaller, as is the land for the use of his family. The school has only 1 ha, where José is allowed to produce for family consumption in exchange of looking after its security.

Even in this limited context, Andrés knows how and what to feed the pigs, deer and other woodland or farm animals that they may raise at different times during the year. *Mbyà* children seem to have more relative competences on knowledge of the woodland compared to the information that peasant children handle, even though their knowledge in this area is still significant. Asked about the plants they know, one of the first mentioned by Andrés and his siblings is the *güembé* (*Philodendron selloum*). They know it grows on other trees or on the ground, and has long roots, which they call *guembepy* and that it has textile uses. Moreover, they know that the fruit, which is 15 or 20 cm long and yellow coloured when ripe, is edible.

Güembé is sought according to the needs of craft production. In terms of communities of practice, children follow the adults and learn from them where and how to find this plant, as well as two bamboo species that are used along with the *güembepy*: the *takuapy* (*Merostachys claussenii*), used as a base for textiles and traditionally used for building walls, among other purposes and the *takuarembó* (*Chusquea ramosissima*), used for braiding.

As we explained above, the access of *Mbyà* communities to the native forest is limited, especially in the studied area where *yerba mate* has been produced for decades. Thus, we could hypothesise that the community of practice composed of adults and children, that enters the woodland searching for these plants, is maintained throughout time, in this case, not because the children cannot find and extract the plants, but because they must cover large distances in order to obtain them.

For Andrés and his siblings, Paola and Hernán, *güembé* is not difficult to identify: in the first place and on simple observation, it is a plant with large leaves of singular shape and its aerial roots are easily recognised. In the second place, it is a widespread plant of the area, with multiple uses, which include ritual ones. Its fruit is used in children's naming ceremonies and also has medicinal uses (basically to eliminate parasites).

Andrés and his siblings do not refer to these ritual and medicinal uses, which are specific patrimony of adult knowledge. However, children know how to use *güembepy* for the fabrication of ropes and animal traps (*simbras*), as well as for the main use that these plants have nowadays: the craft production of basketry, bracelets and rings sold to tourists. It must be noted that even if José and his children do not carry out this activity intensively, their home is located 5 km away from the best known Jesuit ruins in Argentina (San Ignacio Miní), an international tourist destination where many *Mbyà* families of the area sell their crafts in a small nearby market.

Therefore, although Andrés and Paola do not search for *güembé* on a daily basis, they know that the most coveted specimens are found in tall trees, where their roots are longer and more can be made out of them. On the other hand, watching the children handle the *güembé* (Figure 3), we see that they know how to make cross section cuts in order to separate the bark, and throw out the internal soft part of the

Figure 3. Guembe plant in the schoolyard and Andres separating the bark.

plant. After this, they dry out the bark in the sun for a few hours, scrape it in order to even it out and, finally, weave it.

Woodland fruits such as *güembé, guayuvira* (*Campomanesia xanthocarpa*), *aratiku* (*Annona squamosa*), *apepu* (*Citrus aurantium*) or *pakuri* (*Rheedia brasiliensis*), call the attention of Andrés and his siblings, not only because they enjoy sweet food, but also for the interest they arise in other living creatures, especially birds. They are widespread in the area and some of them can be trapped when they are young, fed by the family and tamed, and children can get to know them better. After an activity on birds previously carried out at school with children age 10 or more, I suggested Andrés, Paola and their brothers to look at an illustrated encyclopaedia (Canevari et al. 1991) in order to find out which species they could identify.

Paola immediately identified the *arapachay*, the generic term her father translated into Spanish as *lorito* and *periquito* (small parrot). After looking at four pages with 27 species of the Psittacidae Family (parrots, parakeets, macaws) Paola pointed at two of them: reddish-bellied parakeet or *Pyrrhura frontalis* and the blue-winged parrotlet or *Forpus xanthopterygius*. Through a narration, the child and her brother made a connection between the illustration they were observing and their previous experiences, pointing out that their uncle had recently trapped an *arapachay* that was nesting on a very high tree. By climbing it, he captured a young bird, which they then keep in the community. The children know that these birds can be easily found following the intensity of the young screeching for food, and that they eat maize, tangerines, worms and locusts.

If the *arapachay* is trapped to be raised, the children can identify other birds caught as food, such as the *yerutí*. Going through the Columbidae Family (pigeons and doves), where the book showed 25 different species, Paola immediately identified the eared dove (or *Zenaida auriculata*). Andrés and his younger siblings Paola and Hernán showed how small traps are built with twigs that are placed on the ground in spots where they can see a large movement of leaves that indicates a place frequently visited by birds in search for food. Communities of practices allow the children to develop a precise observation technique that detects traces of birds; identify shapes of leaves and patterns of colours in feathers. This is the way in which these children also immediately identified the *akaé* (jay) as a bird trapped to be eaten. The encyclopaedia presented only three species belonging to the *Corvidae* Family, where the plush-crested jay (*Cyanocorax chrysops*) is differentiated from the rest because of its black head, neck and chest, its yellow iris and its cream-coloured bottom parts and end of its tail – while the other species are completely purple or blue.

5. Conclusion

Our analysis suggests that, even though they share the same school place in a social context at first perceived as homogeneous in terms of the structural position of families, children of peasant and indigenous families have significantly different formative experiences, according to adults' opportunities to carry out different productive activities and thus incorporate young generations in legitimate peripheral participation (Lave and Wenger 1991).

In the first place, peasant and indigenous families own or use heterogeneous parcels of land (between 1 and 200 ha); this significantly determines household production in scope and ability to diversify. Regarding formative experiences and knowledge about the natural and social world, even though most families carry out productive activities mainly for household consumption or for exchange, they currently make efforts to diversify their production through horticulture, animal raising, apiculture and, in larger properties, forestation. This diversity of experiences is very meaningful in the sense that children interact with different animal and plant species, even when access to the native forest is closed to them or is receding.

Furthermore, children participate differentially in productive activities, according to their position in relation to their siblings, their gender and to the main occupation of the father. Contrary to common-sense assumptions, girls also have broad knowledge of plants and animals, although their insertion is subordinate and their experiences are more limited.

Participation in communities of practice (Wenger 2001) of *Mbyà* children is increasingly limited. In the context, we have analysed, it refers particularly to forest extractive activities, involving knowledge about birds and vegetable species for consumption, medicine and craft production. Apprenticeship related with gathering plants involves recognition of species, ability to extract (methods, time and frequency) and elaborate them (treatments for illness, nutrition of human and domestic animals, weave and carving work). Young people can share these activities with adults but not as everyday activities, so a process of appropriation of cultural knowledge is substantively reduced. In the case of peasants, it is important to note that gathering is also a productive activity, but processes of identification tend to link forest with indigenous people, so these practices are less recognised.

Although loss of local knowledge has been reported by academy, civil society and acknowledged by national states in almost all countries, it is relevant to emphasise also the potential for educational contexts when children can reflect on productive practices in the household, challenging the common sense that assumes an alleged link between environmentalism and indigenism, peasantry and conservationism. Social discourse tends to idealise the protection of the native forest by indigenous people, and the techniques of peasantry as inherently sustainable. This naive representation of the relation between humans and nature therefore contrasts with the practices of consumption of plants and animals of children, which allow skilled thought (Ellen 2004) if they could inscribe, in school, their experiences related to different species of plants and animals. These processes might improve lives of peasant and indigenous people, not only because of technical skills that eventually may evolve, but also because of recognition of their knowledge and capacities as subjects and communities.

Science classes can take from everyday children's experiences the suspension of beliefs, searching for alternative explanations of processes when they locate an everyday interrogation in a broader context of relations. When establishing causalities of good nutrition, thinking in terms of networks of understanding that relate pigs with cassavas, when using experimentation to find a better way to trap a bird and, finally, when constructing narratives of these findings, the children are producing knowledge: these ideas can be questioned by schoolteachers not to discredit, but to improve them by lexical codification of knowledge, its textualisation, inscription and formalisation.

Acknowledging the potential of building on what children know, while relating local and scientific knowledge is not new in pedagogical terms. However, this does not take place as frequently or as intensely as expected. Perhaps this means that it is still a challenge to recognise and retrieve knowledge derived from social practices, which is not formalised nor expressed as such. This article aims to contribute in that direction, by systematising knowledge that children from peasant and indigenous communities from the Northeast of Argentina have constructed in their out-of-school daily practices.

Acknowledgements

This research is made possible by financial support from the University of Buenos Aires and the National Council for Scientific and Technical Research (CONICET), from Argentina. I thank Mercedes Pico for her work in translating this article, the guest editors of this special issue and the anonymous reviewers for their useful comments. Some ideas where discussed at Oxford's Ethnography of Education Conference (OEC) in 2011, I will also thank relevant comments received there.

Notes

1. Current archaeological research in Misiones shows sites of up to 1200 BP near Yaciretá, and 900 BP in Oberá, even though occupation came as far as the Iberá Marshes and the islands at the mouth of the Paraná river. Dates in nearby Rio Grande do Sul (Brazil) suggest that it is very likely that in Misiones and Corrientes older data could be found, and that early occupation goes as far back as 2000 BP (Noelli 2004, 31–2).
2. Bartolomé points out that these terms are external classifications applied to different ethnic organisational groups belonging to one ethnolinguistic group. This is not a typological problem of mere academic interest; it represents one of the keys for accessing ethnic identifications of members. The term 'Guaraní' applied to refer to them became widespread only in the seventeenth century, when as a result of the type of cultural collectivity constructed by contemporary states, ethnic groups were thought of as totalities in national terms, with the implicit assumption that an ethnic group corresponds to a 'nation', understood as a social community with conscience of self and able to set common goals. However, the 'nation' is a type of collectivity historically constructed by States as of the nineteenth century, which set to develop politically and culturally homogenous social communities. The Guaraní tradition developed without the presence of a broad state formation, that is, without a political apparatus that could unify the different populations and diverse linguistic varieties (Bartolomé 2004).
3. The concept of squatters involves significant diversity. While a minor sector shares many common features with the farmers, most can be defined as peasants who are scarcely integrated into the economic system, and a number of them are semi proletariats that combine salaried work with minimal subsistence agriculture (Baranger 2008).
4. The concept of farmer has been much discussed in the anthropology of this region for many decades. The work of Archetti and Stolen in the North of Santa Fe (1975) and that of Bartolomé in Misiones (1975) discuss the non peasant character of the agricultural producers of the area, using the term *farmer* – on occasions translated as '*granjero*'. This form of production, characterised by the use of domestic labour but, whenever possible, a certain accumulation of capital, lead the farmers to focus their political claims on the access to credit, commercialisation and merchandise distribution (Baranger 2008).
5. Although the names of towns and communities have been maintained with the objective of contributing to the region's economic and social documentation, the children's names have been modified in order to preserve their anonymity. It is important to notice that *Mbyà* children have names in *Guaraní* but they use a name in Spanish for school.

References

Archetti, E., and K.A. Stolen. 1975. *Explotación familiar y acumulación de capital en el campo Argentino* [Family production and capital accumulation in the Argentinean countryside]. Buenos Aires: Siglo XXI.

Baranger, D. 2008. La construcción del campesinado en Misiones: de las Ligas Agrarias a los "sin tierra" [The construction of peasantry in Misiones: from Agrarian Leagues to "without land people"]. In *Campesinos y agricultores familiares. La cuestión agraria en Misiones a fines del siglo the twentieth* [Peasants and farmers. The agrarian issue in Misiones at the end of the twentieth century], ed. G. Schiavoni, 33–70. Buenos Aires: Ciccus.

Bartolomé, L. 1975. Colonos, plantadores y agroindustrias. La explotación agrícola familiar en el sudeste de Misiones [Colonos, farmers and agribusiness. Family plots in the south-east of Misiones]. *Desarrollo Económico* 58: 25–56.

Bartolomé, L. 2000. *Los colonos de Apóstoles: estrategias adaptativas y etnicidad en una colonia eslava en Misiones* [Colonos from Apostoles: Adaptative strategies and ethnicity in an eslavonic colonia in Misiones]. Posadas: Editorial Universitaria, Universidad Nacional de Misiones.

Bartolomé, M. 2004. *Flechadores de jornales. Identidad guaraní en el Paraguay contemporáneo* [Day's Pay Archers. Guarani's identity in contemporary Paraguay]. Amérique Latine Histoire et Mémoire. Les Cahiers ALHIM (online February 2, 2005). http://alhim.revues.org/index120.html (accessed May 27, 2009).

Bidaseca, K., and D. Mariotti. 2001. Viejos e nuevos actores en la protesta rural en la Argentina [New and old actors in rural protest in Argentina]. *Sociologías* 3, no. 5: 148–89.

Canevari, M., G.R. Carrizo, G. Harris, J. Rodríguez Mata, and R. Straneck. 1991. *Nueva Guía de las Aves Argentinas* [New guide of Argentinean birds]. Buenos Aires: Fundación Acindar.

Cebolla Badie, M. 2005. Ta'y ñemboarái. La miel en la cultura mbya-guaraní [Ta'y ñemboarái. Honey in mbya-guarani culture]. Programa de Doctorat en Antropologia Social i Cultural. Universitat de Barcelona. *Resumen de Trabajo de Investigación Bienio 2002–04.*

Centro de Trabalho Indigenista (CTI). 2008. *Guarani Reta. Pueblos Guaraníes en las fronteras de Argentina, Brasil y Paraguay.* Brasília: Centro de Trabalho Indigenista.

Cragnolino, E. 2006. Infancia, escolarización y apropiación de la lectura y la escritura en familias rurales [Childhood, schooling and appropriatrion of reading and writting in rural families]. Paper presented at XI Simposio de Investigación Etnográfica en Educación, March 20–24, in Buenos Aires, Argentina.

Ellen, R. 2004. From ethno-science to science, or what the indigenous knowledge debate tells us about how scientists define their project. *Journal of Cognition and Culture* 4, no. 3: 409–50.

Gobierno de Misiones. 2008. *Anuario Estadístico* [Statistical yearbook]. Misiones: Instituto Provincial de Estadísticas y Censos.

Gorosito, A. 2005. Liderazgos guaraníes, breve revisión histórica y nuevas notas sobre la cuestión [Guaranies leaderships, historical review and new notes about the issue]. Paper presented at VI Reunión de Antropología del Mercosur, November 16–18 Montevideo. Uruguay.

Instituto Nacional de Estadística y Censos (INDEC). 2010. *Resultados definitivos de la Encuesta Complementaria de Pueblos Indígenas 2004–2005.* http://www.indec.gov.ar (accessed March 28, 2010).

Jaume, F., C. Villar Gonzales, and Y. Urquiza. 1989. *Notas Sobre la Historia de Misiones: El Proceso de Formación de la Región Histórica* [Notes about the history of Misiones: The formation process of the historical region]. Posadas: PISPAD.

Lave, J., and E. Wenger. 1991. *Situated learning: Legitimate peripheral participation.* Cambridge: Cambridge University Press.

Manzanal, M., and A. Rofman. 1989. *Las economías regionales de la Argentina. Crisis y políticas de desarrollo* [Regional economies of Argentina. Crisis and development policies]. Buenos Aires: CEAL.

Noelli, F.S. 2004. La distribución geográfica de las evidencias arqueológicas guaraní [Geographical distribution of guarani's archaelogical evidences]. *Revista de Indias* LXIV, no. 230: 17–34.

Otero, N. 2008. De la chacra al corte. Los dirigentes de Unión Campesina y la lucha por la tierra en el nordeste de Misiones [From farm to protest. Peasant's Union leaders and the struggle for land in the north-east of Misiones]. In *Campesinos y agricultores familiares. La cuestión agraria en Misiones a fines del siglo XX* [Peasants and farmers.

The agrarian issue in Misiones at the end of the twentieth century], ed. G. Schiavoni, 71–98. Buenos Aires: Ciccus.

Reboratti, C. 1979. Migraciones y frontera agraria: Argentina y Brasil en la cuenca del Alto Paraná-Uruguay [Migrations and agrarian frontier: Argentina and Brasil in the high Parana-Uruguay basin]. *Desarrollo Económico* 19, no. 74: 189–209.

Ricotto, A., and J. Almeida. 2002. Las ferias francas de Misiones, Argentina: una red de actores sociales y una nueva visión del mundo rural [Free markets in Misiones, Argentina: A network of social actors and a new vision of rural world]. Paper presented at VI Congreso de la Asociación Latinoamericana de Sociología Rural, Porto Alegre. http://www.ufrgs.br/pgdr/textosabertos/artigo%20ricotto.pdf.

Rockwell, E. 1995. *La escuela cotidiana* [Everyday school]. México: FCE.

Rogoff, B., J. Mistry, A. Göncü, and C. Mosier. 1993. Guided participation in cultural activity by toddlers and caregivers. *Monographs of the Society for Research in Child Development* 58, no. 8: 1–179.

Schiavoni, G. 2008. Nuevas organizaciones agrarias. Plantadores y campesinos en el nordeste de Misiones [New agrarian organizations. Farmers and peasants in the north-east of Misiones]. In *Campesinos y agricultores familiares. La cuestión agraria en Misiones a fines del siglo XX* [Peasants and farmers. The agrarian issue in Misiones at the end of the twentieth century], ed. G. Schiavoni. Buenos Aires: Ciccus.

Sili, M. 2005. *La Argentina rural. De la crisis de la modernización agraria a la construcción de un nuevo paradigma de desarrollo de los territorios rurales* [Rural Argentina. From crisis of agrarian modernization to the construction of a new paradigm of development in rural territories]. Buenos Aires: Ediciones INTA.

Wenger, E. 2001. *Comunidades de Práctica* [Communities of practice]. Barcelona: Paidos.

Wilde, G. 2007. De la depredación a la conservación. Génesis y evolución del discurso hegemónico sobre la selva misionera y sus habitantes [From depredation to conservation. Genesis and evolution of hegemonic discourse about Misiones' forest and its inhabitants]. *Ambiente y Sociedade* X, no. 1: 87–106.

Collaborative imaginaries and multi-sited ethnography: space-time dimensions of engagement in an afterschool science programme for girls

Jrene Rahm

Département de psychopédagogie et d'andragogie, Université de Montréal, Montréal, Canada

Temporal and spatial configurations that constitute learning and identity work across practices have been little explored in studies of science literacy development. Grounded in multi-sited ethnography, this paper explores diverse girls' engagement with and identity work in science locally, inside a newsletter activity in an afterschool programme and globally, by exploring two girls' trajectories of identification and cultural repertoires of science over time and across space. This study illustrates the ways diverse space-time configurations marked the girls' local meaning making of science. The study also speaks to the importance of recognising the diversity of girls' trajectories and positioning in science and the difficulty in understanding their layers of contradictions. Essentially, the collaborative imaginary is a partial account and co-construction of researcher and researched that unpacks the many time-space configurations that mark diverse girls' science literacy development in a world characterised by globalisation.

Introduction

> Myself as a scientist: Assume that you are grown up and work as a scientist. You are free to do research that you find important and interesting. Write some sentences about what you would like to do as a researcher and why.
>
> Mohini responded by email: 'I would like to work at NASA, and pursue research on extra-terrestrial life, to pursue research on intelligent life elsewhere. The reason being for instance, on Mars, there was some life before us, so that planet which is not the Earth, had the right conditions at one point to support life. So, evidently, there could be life in outer space somewhere at this very moment.
>
> Also, if I would find life somewhere, I would like to communicate with these people right away . . . or then, I would like to pursue research on the Martial colony that will be developed in many years. For instance, find a way to increase the density of the Martian atmosphere in outer space'. (Selection of text from email-exchange with youth, Summer 2010)

Mohini received the above assignment a week after participation in a summer activity with a local scientist and in collaboration with a classroom in France, building a model of an imaginary life-bearing planet. The activity took place in an afterschool programme in which she had participated for many years. Her response underlines some of the methodological and conceptual challenges studies of science literacy development raise and that will be addressed in this paper. In essence, Mohini's narrative of scientific research is strongly grounded in a project she just completed rather than a reflection of a critical integration of her stance and interest in science. It differs significantly from her interview talk about science and self in science that was collected earlier where she voiced her interest in research on ecology and plant biology. It leads to questions about the manner time and space mark youths' science literacy development and identity work. What would have been her answer a week prior to that experience, or two years from now? In other words, in what ways does time and space constitute the manner she understands science and how she positions herself within science? How can we take time-space configurations into consideration in our studies of science literacy development?

These are some of the questions I attempt to respond to in this paper through stories of girls' science literacy development and identity work in science in an afterschool programme. I begin with a description of the theoretical grounding of science literacy development in sociocultural historical theory and discuss its implications for learning, identity work and its temporal and spatial ramifications. I then offer a description of the research site and its participants before exploring local engagement with and identity work in science in the newsletter activity. Next, I move on towards the global and present trajectories of identification as embedded in complex time-space configurations. I conclude with suggestions for future research and implications for policy.

Science literacy development: learning, identity and time-space configurations

Grounded in sociocultural historical theory, learning science and becoming an insider to science is understood in this paper as constituted by engagement in and with science in cultural practices that are valued by its participants – not simply because of what can be learned about science and the ways of being in science that are made possible. Instead, participation in science is driven by the social value these new ways of knowing and being may have for participants within and beyond that practice (Gutiérrez and Lee 2009). Culture in itself is understood as dynamic and hence, as a verb and action, activity or practice rather than trait of an individual such as an individual's 'ethnicity' or 'race' (218). Accordingly, youths' science literacy development is taken as constituted by repertoires of cultural practices in which youth participate in over time, and across space, and that constitute their lives. In the words of Gutiérrez and Lee (2009, 28):

> Learning is oriented toward shared practices that are historically and locally situated, and organized around students' repertoires of practices – that is, practices developed in the movement and flow as youth move across everyday settings and practices.

In line with network theory, science literacy development has to be understood in terms of its ways of producing space and time, yet also movement across space and time, supportive of 'fluid and contested' identities and forms of engagement with science, 'that are simultaneously also frameworks of power' (Nespor 1994, 9). It has led to a focus on what Nespor (2004) refers to as educational scale-making, given the assumption that 'teaching and learning play out in complex and layered processes in which actions and events are simultaneously produced and made meaningful at multiple spatio-temporal scales' (309). Put differently, the development of science literacy needs to be grounded in studies that take seriously the 'simultaneity of multiple locales' and the manner in which the locales overlap (Leander, Phillips, and Taylor 2010), yet also the timescales of action, social practice and activity (Lemke 2000). In that sense, studies of the local always imply the global and hence, need to focus on 'youth moves' and how materiality yet also positioning make certain movements possible, yet not others (Dolby and Rizvi 2008).

Building on these ideas in this paper, I explore science literacy development and identity work in science as embedded in a complex web of spatial dimensions and as constituted by different timescales. The former entails the mapping out of spaces of science literacy in the making and youths' mobilities within and among such spaces, while the latter is explored through a focus on learning and identity trajectories within and across spaces over time (Leander et al. 2010). Together, they make up a collaborative imaginary of science literacy development grounded in sociocultural theory and extensions thereof (Gutiérrez and Rogoff 2003; Lemke 2000), and that also draws upon the literature on network theory (Nespor 1994, 2004), and spatial theory in geography (Harvey 1990; Soja 1996; Vadeboncoeur 2009). I use collaborative imaginary to emphasise its grounding in multi-sited ethnography, as developed by Marcus (1998), in which stories that emerge from research are understood as co-constructions among the researcher and researched, and always partial.

Science literacy in itself is understood as multi-dimensional, entailing the appropriation of critical ways of thinking, doing and talking science that are seen as embedded in a complex socio-political landscape and system. Scientifically literate youth are the ones that see themselves inside science and as critical actors who can put science to use for social action and in ways that are empowering to them (Basu, Calabrese Barton, and Tan 2010; Bell et al. 2009). Studies exploring science literacy development have become focused on examining what identities in science are available to youth through participation in an array of cultural practices that are part of historical cultural constellations of science practices in their communities (Gutiérrez and Rogoff 2003). Yet, it is also about an exploration of movement in a network of practices in which school science – dominated by a 'pipeline' ideology with a strong focus on cookbook labs and talk about science, but little meaningful engagement with science – is the most stable and well-established practice, against which other practices of science have to compete (Aikenhead 2006). Given the position of school science, it is clear that actor-networks, practices and emergent identities and alliances with science are about power (Nespor 1994).

Building upon the work of Holland et al. (1998), it is essential to understand 'how people come to understand themselves, how they come to "figure" who they are, through the "worlds" they participate in and how they relate to others within and outside of these worlds' (Urrieta 2007, 107). While the social positioning of youth happens through media, everyday discourses, spatial arrangements, texts and other meditational means in activity (Holland and Leander 2004), the youth author themselves in relation to those ascribed social positions. They actively engage in rejecting, accepting or negotiating the positions either assigned or offered to them, as they move within networks of practices, leading to new possibilities of being or becoming. The contesting of social positioning and self-authoring in a local practice is the focus of this paper, yet it is simultaneously understood within the global and youths' histories of movement among practices (Holland and Lave 2001; Holland et al. 1998).

The making of identities has to be understood as constituted by forms of engagement in practice. Accordingly, the study focuses on the ways girls engage with science, the affordances and constraints that constitute the doing and knowing of science as well as ways of being in science. Simultaneously, the study explores youths' trajectories of learning and identification within and beyond the programme ScienceGirls. It is assumed that girls' trajectories take 'form in the flow of historically, socially, culturally, and materially shaped lives' (Holland et al. 1998, 5). Thus, the study explores in what ways the girls authored a sense of self, yet also potentially transformed the local into a practice of agency thereby challenging its transgression and reproduction of forms of engagement and identities in science they already knew from elsewhere (Holland et al. 1998).

Grounding this imaginary in multi-sited ethnography made possible a journey into the spatio-temporal scales of science and identity in science in out-of-school time (OST) programmes. As suggested by Marcus (1998), multi-sited ethnographies bring together space and time in ways 'previously thought incommensurate' (14), in that they are:

> . . . designed around chains, paths, threads, conjunctions, or juxtapositions of locations in which the ethnographer establishes some form of literal, physical presence, with an explicit, posited logic of association or connection among sites that in fact defines the argument of the ethnography. (Marcus 1998, 90)

Accordingly, the bringing together of spaces and moments not only happens by listening in on subjects' talk about their trajectories over time and across space but also implies travel with subjects, and hence, the creation of 'collective aspects of research' (Marcus 2009, 186). In the study presented here, the *thing* – science literacy as constituted by repertoires of practices – held the sites and study together. By following its life and its biography, I explore different forms of engagement and identity work in science among a group of first- and second-generation immigrant girls. By referring to such research endeavours as imaginaries, Marcus (1998) also purposefully hints at their inherent dialectic and hence, co-construction of the story among the researcher and researched and the kind of positioning it implies of them both. It is literally a story that unfolds and emerges as the researcher and researched together construct what science literacy development as constituted by repertoires of cultural practices means. Essentially, the qualitative research imaginary makes possible the delving 'into the vibrancy of the "in-between" in important ways, in that it recognizes that education unfolds across numerous sites and settings with and between multiple texts such as the Internet, music, dance, and other aesthetic spaces' (Weis and Dimitriadis 2008, 2307). It is a method of choice for the study of 'the conditions of changing world systems' (Pierides 2010, 179), able to capture 'emergent notions of time and space' (181), and thereby moves the study of science literacy development beyond static notions of 'ethnos, culture or society' (179). Thus, I illustrate that moving through co-constructed stories that depict what it means to engage in science and develop identities in science.

Overview of the research imaginary

I draw from a multi-sited ethnography grounded in two community-based organisations (CBOs). In this paper I focus on ScienceGirls, an afterschool programme that has been in existence since 1987, serving diverse urban girls from two elementary schools in an underserved community of Montreal. The programme is housed in a community centre offering diverse services and spatially close to the two elementary schools that it serves. The goal of the programme is to offer girls with an array of science activities that encourage the development of the tools needed to complete high school and to pursue higher education.

The scientific newsletter activity is the focus of a three-year video ethnography (2008–2011), supplemented by individual and focus group interviews of the girls and their instructors. I draw upon data collected primarily in the first two years of the study. Twelve girls participated in the activity each year ranging in age from 11 to 14 years (6th grade in elementary school and 1st and 2nd year high school), girls with a mixed immigration history, with parents from Morocco, Congo, Caribbean, Greater Antilles, Bangladesh and Sri Lanka. Two youth assistants (Alana and Mohini, who will be introduced later), Danisha, the activity coordinator, and two other adults, supported the writing of the newsletter. The making of two newsletters – Music and Nature – was examined during the first year of the study (January to May 2009), and five newsletters in the second year (Fall 2009 to June 2010): School and Science; the Universe; Biodiversity; Human Body; and Oceans and Continents. The newsletters were distributed in the two elementary schools in its vicinity and through the CBOs website. In line with the language act in place in Quebec since 1977 that proclaimed French as the official language of that province, all programme activities were conducted in French and all youth attended schools conducted in French. However, code switching among French, English, Tamil and Hindi was common. The stories in this paper are a bricolage (Kincheloe and Berry 2004) of some of the many diverse data sources that were collected over the two-year study.

Complicity of research imaginary

In the current study, the affinity between the fieldworker and researched was marked by shared histories in terms of their identity and trajectory as immigrants, as having had little access to science in more formal institutions or at home. Yet, we both were interested in science and curious about learning more about science, recognising its pertinence to daily life in the face of new challenges. That complicity, which stands for a kind of partnership or shared endeavour that made us work together, constituted in important ways the making of the research imaginary and understanding of the girls' repertoires of cultural practices of science. It also constituted the girls' identity work in and beyond science. I briefly introduce the personal history of the two focal girls in this paper, next to my own.

Alana was born in Montreal and is now 15 years old. Her mother is from Trinidad and arrived in Montreal at a young age. Her father is from Saint Lucia and arrived when he was about 16 years old. He is bilingual and finished a college education. Her mother pursued a degree in economics at one of the local universities but never completed it. Alana's mother owns a restaurant together with her mother, specialising in Trinidadian cuisine. Alana has a younger brother who is 10 years old. Alana goes to a French public high school and has always been in the academically enriched class. Alana participated in ScienceGirls since Grade 4. While she moved away from that neighbourhood, she still returns Fridays to work as an assistant. She writes the editorial of each newsletter at home, and helps others with their texts while at the programme.

Mohini, whom you met in the Introduction section, was born in Lugano in Switzerland and is now 14 years old. Both her parents are from Sri Lanka. Her father managed to secure a job as a cook in a restaurant in Switzerland, making the whole family leave their home country. But her mother could not work in Switzerland and eventually decided to move on to Canada to join some of her family that had emigrated there earlier. Mohini's father remained in Switzerland where he still works as a cook. Mohini has a younger brother who was born in Canada. They speak Tamil at home while her mother is also trilingual like her, speaking French and English. Her mother wanted to teach mathematics again in Canada but was asked to redo her teacher certification at university. She eventually opted for a college degree as a pharmaceutical technician. Mohini finished her second year of high school in a girls-only public school in the enriched track. Her mother also pursued an education in a girls-only school in her home country. Mohini participated in ScienceGirls since 5th grade and enjoyed most the science fair projects.

Jrene, the researcher, like the girls, is also an immigrant with parents who lack a university education. Like them, I am multilingual, having grown up in Switzerland, speaking Swiss and German primarily, and later having learned English, and picked up on whatever French was left from high school as I started my work as a professor in Montreal in a French university. Alana and Mohini are fluent in English and French and switch forth and back without hesitation, which sets me apart from them. Next to my trajectory as an immigrant, my history in science as a first-generation college bound woman is another dimension of our affinity that came to constitute the research imaginary in important ways.

A collaborative imaginary of the local: the scientific newsletter in the making

> One girl is busy on the computer, cutting some images of 'Dora the Explorer' into a word document. She then talks to me about a movie from India that she watched at home the other day and takes me to a website with its soundtrack and plays the part of it that is available on-line. Other girls arrive and start discussing the American TV Show 'How to be Indian' which prompted them to google bollywood.com for me.
>
> Danisha, the lead adult instructor of the newsletter activity, calls the girls to the big table for the snack and whole group discussion period prior to individual work on the newsletter. It is time to vote on the next newsletter theme. Danisha announced that they voted for the theme Feelings prior to their Christmas vacation since their favourite topic, Paranormal, was judged as not scientific enough. Her announcement caused an immediate uproar among the nine girls now sitting around the table. The two adults and two youth assistants encourage the girls to come up with new topics, 'as long as they are scientific'.
>
> They pulled up a new page on the poster board to note the new themes. One girl suggests, 'Fear', another 'Year 2012', others add themes that were on the list previously, such as writing about their home country. The latter is quickly dismissed as not scientific enough. Danisha suggests the theme 'Weather' and refers to the freeze in Florida that was happening at the time, mentions Tsunami, global warming and natural disasters. One assistant gathers some magazines in the adjacent room that cover natural disasters to offer

> some concrete ideas to think with. Eventually, they end up with the theme 'Biodiversity' since some girls were interested in the study of animals in danger of extinction.
>
> Now, the kind of article to be written needs to be chosen by each girl or team. One girl opts for an article on an animal and brainstorms together with Alana, one of the youth assistants, who lists 'dogs, cats, giraffe' to get her started. Yet, she wants to talk of an animal from India, her country of origin, and Alana continues with a new list 'tiger, elephant...' making the girl opts for the elephant. Together, they brainstorm about some of its symbolic meanings in India such as the trunk being high up in the air bringing good luck. Later, they discuss different ways to approach a scientific text on elephants, talking about its threats for survival, differences due to its place of origin, such as those from Asia as opposed to those from Africa. (Condensed Fieldnotes, January 8, 2010)

As the above description underlines, engagement with science at ScienceGirls entailed exploring science content of interest to and chosen by the girls beginning with one of their questions and eventually leading to a text in which that science content is accurately presented and explained. The fact that paranormal was judged as non-scientific by the staff suggests that scientific content was key, rather than scientific reasoning which could have taken the form of wondering about why paranormal may or may not be scientific. The excerpt suggests that participation in the programme entails foremost an appropriation of what counts as science in this space. It is evident in the kind of guidance one of the girls receives as she struggles with a scientific text on elephants. She is encouraged to move beyond the symbolic meaning of that animal in her home country and her personal interpretation of it also marked by Hinduism, since that was considered inappropriate for the newsletter. It demonstrates the narrowly defined, constitutive notion of science used in the newsletter. The example suggests that even though youth agency in terms of topic choice was encouraged, there were some limits, given the instructors' narrow representations of science and youths' own history and understandings of science. It suggests complex spatial overlap that marked engagement with science.

Simultaneously, the narrative makes evident the cultural flow and disjunctures, yet also the interspatiality of meaning making and identity work. For instance, the author of the story about elephants positions herself simultaneously as the culturally knowledgeable girl from India, as somebody who enjoys connecting her life and ethnic self with science. She also presents herself as techno-savvy; and as needing guidance by her older peers and adults in the programme. Her positioning work clearly takes us places. It shows in what ways her history as a second-generation immigrant from India marks her present. At the same time, it hides her present academic position as a girl having made it through first year of high school in a special class with others that also struggle academically. She followed the 6th grade curriculum in her first year of high school, in part due to her struggles with French. She never shied away from emphasising how much she valued the opportunity to practice her French in the programme while feeling safe and not being made fun of by anybody, positioning herself as in need of help. She did not shy away from an activity that entailed writing in French but instead perceived it as an opportunity to grow and valued the public recognition and attention the newsletter issues received given their distribution on the web, in the community and schools: 'It has my name on it, and so it is clear that I wrote the article and to show everybody, 'look, it is me who wrote this and it's good'! This activity positioned her as capable despite her academic struggles in school. She tried hard to re-position herself that way, and to convince others of her position as capable and as becoming educated.

Engagement with and identity work in science was also mediated by diverse cultural tools, one of them entailing a participation structure often found in OST settings (Deutsch 2008), where older youth like Alana take on the role as youth assistants, guiding youths' writing of texts for the journal, while they are simultaneously mentored by the adult instructors in the programme. Note Alana's crucial role as a youth assistant in guiding the girls' (Ajala and Fafa) framing of an article below when talking with an adult instructor (Naomi):

Alana	So what are your questions now?
Ajala	Why has chocolate been invented?
Alana	But it's delicious . . ..
Ajala	No, no, no, why did we invent it, because it tastes good, but how does one make chocolate, that would be a better question
Alana	Yeah, ok, how does one fabricate chocolate, that also came to my mind
Ajala	Or, also, why do we call it chocolate, why is it called that way?
Naomi	But girls, do not get too far off the theme – the theme of the newsletter is about the human body . . .
Ajala	So then, how does the body taste the taste of chocolate?
Alana	Wait, you need to rephrase it, that does not really work (later) Maybe we could say, what are the kinds of tastes that the tongue is able to distinguish
Fafa	Yeah, the salty, the sweet, . . .
Alana	No, no, let's see, with chocolate, there is the bitter, that's it, if you got black chocolate, it is bitter
Fafa	Ah, it's disgusting
Ajala	I know!
Alana	But I find it good. You, it's because you're young, and so you got the taste for sugar, you just need it, I too never liked it before, but now, I like it.
Ajala	Right, it's because you change, your body changes.

Initially, Alana offers some suggestions, and then helps the girls re-formulate the question in ways it aligns well with a scientific newsletter. It made Alana propose a link between taste and the function of the tongue. It moved the topic towards science while the girls challenged that move by connecting their discussion back to personal experiences about their liking or disliking the somewhat bitter taste of dark chocolate. They manage to tie their conversation back to the topic of the newsletter by noting how ones' taste may change due to changes in the human body.

The analysis points to a complex interweaving of many worlds and ways of knowing that engagement in science in the newsletter activity makes possible. While not all ways of talking science are accepted, they are respected resources and put to use. The collaborative imaginary underlines the ways in which the study of meaning making of science took me places that I needed to better understand as a researcher if I wanted to figure out the role engagement in science in this space played in the lives of the participating girls. I had to learn about Ajala's country of origin and culture that constituted her actions in the programme (Hinduism, Bollywood, etc.), her interest in paranormal as tied to her cultural roots, her position in school and her educational aspirations, her perception of self within and beyond the programme, ad within her family. Yet, I struggled in my attempts to attend to my own stereotypical visions (e.g. about Bollywood, Hinduism and what it means to be South Asian). They intersected with what I saw and the youths' enactment and talk about self that often appeared contradictory and was clearly a complex cultural hybrid of their history and trajectories yet also their media image of their home culture (Handa 2003).

Time-space configurations and trajectories of identification

By weaving together talk about forms of participation and identity work in science as lived by Alana, Mohini and myself, I use our affinity as a tool for deconstructing girls' engagement with and positioning in science in yet another manner.

Alana's trajectory

The designated identity of the newsletter activity in ScienceGirls entailed the becoming of a journalist, an identity marker the participating girls embraced. It helped the girls to position themselves at the margins of science, something they felt comfortable with. Alana's quote is telling:

> I think it is important to inform others about science, it is important to understand science. I like to read about different kinds of sciences and technology. And I think all sciences, it is important, it helps us understand ourselves, our environment, the society in which we live and the human sciences. I do not have any passion for the pure sciences, mathematics, chemistry, physics, I prefer the human side of things, like sociology, psychology, maybe physiology but even then that is too technical for me. I like thinking about things.

Alana positioned herself as having no passion for the pure sciences but as liking to think about things and highly valuing science as a tool for action – to talk about, to share with others, to think about, to better understand oneself or ones environment with it. Alana claims to enjoy 'thinking about things' which is used almost as an excuse to position herself as interested in sociology or psychology rather than the pure sciences. Her identity as a user of science knowledge to make informed choices also travels places and is spatially marked not only by the newsletter activity but also interactions with science at home:

> My mother used to smoke, she now stopped since three weeks. And so I helped her, I went on-line and used the computer to discover things about the lungs and stuff, to encourage her to continue to stop smoking, despite all the symptoms you get when you stop, when your body does no longer have the nicotine, but after a while your body gets used to it.

Her identity as user of science to take action is embodied in her support of her mother as she quit smoking. Such ways of engaging with science mirrored my own, as I was trying to explain to my father what is happening with his brain due to Parkinson, a disease he struggles with and that made me confront science in ways never pursued before. Science literacy became valuable to both Alana and me, while science remained beyond the reach of our parents, who positioned themselves publicly outside of it. That became clear through a conversation with Alana's mother Shane, about nutrition and traditional cooking at her restaurant. Shane knew much about the traditional ways of Caribbean cuisine (Trinidad) and the manner such ways of cooking and eating exploit different combinations of food to assure vitamin balance and health. Shane's expertise was marked by her history and the cultural food practices that she was enculturated into at the elbows of her mother and founder of the restaurant. Yet, Alana and Shane positioned themselves outside of science given their shared vision of science as entailing the enculturation into the pure sciences only, and 'a body of authoritative, incontestable knowledge which is abstracted from social activity to maintain a high level of difficulty and status' (Hughes 2001, 276).

When exploring Alana's trajectory of identification in and with science over time, next to her repertoire of cultural practices in Figure 1, it becomes evident that she never had access to spaces in which she could have developed more expansive visions of science or ways of being in science. For her, science learning is about 'deference to the experts of science, which implies that scientific knowledge is not constructed among students but delivered to them' (Seiler and Gonsalves 2010, 100).

Referring to school in her last year, she further distanced herself from her vision of the pure sciences by recalling her struggle with mathematics exclaiming, 'I really cannot take math anymore'! She blamed herself, claiming she did not have what it would take to become an insider to science, as her discourse about scientists reveals, 'scientists can be male or female, but they have to be hard working, you have to be studious to be a scientist and organized'. As suggested by Phipps (2007), at the heart of many initiatives to bring girls into science is the notion that girls are lacking the knowledge about science or identities in science and hence, need to be given access to such ways of knowing to then pursue science. Yet, they leave unquestioned deeply grounded images of science and its practices that keep girls out. That contradiction was clearly at work here in that Alana was interested in science but did not believe in having what it takes such as 'being studious and hard working'. Yet, she strongly believed in the importance of science literacy since 'it makes us more critical, curious,

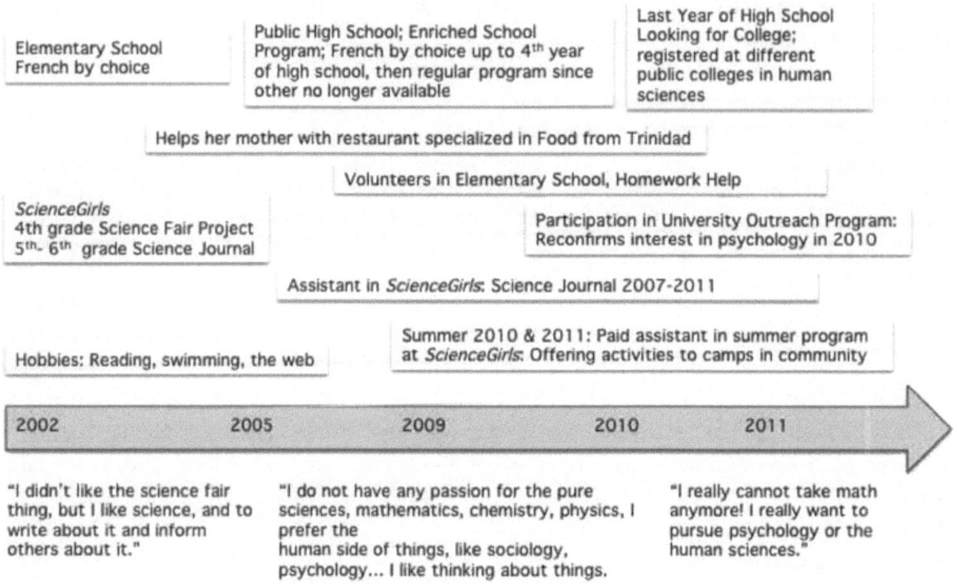

Figure 1. Alana's repertoires of cultural practices and trajectories of identification.

so we try to find out about still more things, I think this is important'. Alana's statement underlines her strong interest in science and an identity as a user of science for action. She could see herself as an insider in the newsletter activity, yet not the science fair activities in ScienceGirls that she abandoned early on, maybe in part because they were not commensurate with her position in science. Accordingly, her trajectory of identification in ScienceGirls over time reinforced in complex ways her vision of self in science and what science is. The newsletter activity made possible her enactment of self as a writer about science and hence journalist that puts science to use by informing others, yet it left unquestioned her pipeline image of science and perception of science as difficult and inaccessible especially now, given her struggle in mathematics. Her local experiences in the newsletter activity did not offer her with the tools needed to re-position her inside science beyond that setting. In light of network theory, the newsletter practice was a fragile practice embedded in a complex web of other practices and discourses that reinforced her positioning outside of science given their power (Nespor 2004).

Mohini's trajectory

Alana's positioning work aligns itself at times with the one also apparent in Mohini's trajectory of self-identification. Yet, Mohini's recent experience and success in a science fair project, winning its gold medal at the local, provincial and trans-Canadian level in the spring of 2011, made her publicly announce for the first time and share her position inside science, as interested in medicine, astrophysics and physics:

> Since the Science Fair experience, I am more open now to say to others, OK, I love science... there will be people that find me strange, but hey, now I can say it to others, that I enjoy science, that I love science...

This statement underscores the social stakes associated with girls' identification as insiders to science. That girls need to hide their interest in pipeline science is not new given historically well-grounded discourses about who gets to be a scientist (Schiebinger 1999). For instance, I learned early on in Switzerland, that science is reserved for boys and the social stakes associated with 'wanting a career' as a woman, even in the social sciences. A report about women in science in Switzerland (Knobloch 2009) offers interesting insights into improvements that have been made, in that 50% of students pursuing higher education are now female (60% in Canada). Yet, their representation is highly skewed by content area with for instance, 85.8% in the applied sciences and only 6.4% in technology. Women are still highly under-represented in higher-level institutions as well. To some degree, it confirms the still rather desperate situation of women's representation in the science pipeline and higher education that I experienced, with many areas of gender bias still highly under-researched. Mohini's description of her brother's engagement with science and position as insider to science sounded too familiar:

> He takes the science books and learns, he is learning about how to calculate the volume of a cylinder, he invented a magnet that makes a door open by itself, he is good, sometimes I am jealous of him...

It was her brother that she described as able to do calculations and as being able to invent things, two dimensions she perceived as essential to an insider identity to science. It resonates with the discourse and positioning work of my brother and myself that marked my growing up and that made me leave my home country in search of a place where I could pursue an education that I valued.

Mohini's interest in science made her seek out ScienceGirls. Unlike Alana, Mohini enjoyed the pursuit of science fair projects. She preferred to stay engaged with science by writing and doing. It made her participate in the science fair projects for three consecutive years while at the elementary level and again this year (2010–2011), in a new pilot project for high school girls with a focus on science fair projects that are then presented at the local science fair competition. When we first met in 2009, Mohini summarised her interest in science as follows:

> I like science and mathematics a lot. I like anything about plants, evolution, I have a lot of plants, and mathematics, my mother was a math teacher in Sri Lanka. . . .
>
> The sciences, I find them very important because without science we could not heal people, there are so many things that we could not do if there was no science, and I like everything, but the thing I like most is evolution, I am not sure how to say it but I find . . . it was really my history teacher that taught me all about the beginning of life, he talked to us for two days about the Big Bang, it was so interesting, I loved it!

Like Alana, Mohini described science as interesting and important for us to know about and put to use. Also like Alana, she distanced herself from the hard sciences, particularly robotics. Mohini valued 'plants and evolution', aligning herself with areas of science in which women figure more prominently (Knobloch 2009; Schiebinger 1999), despite her mothers' encouragement and identity as a math expert, as Mohini called her. In fact, unlike Alana or myself, Mohini had access to role models in her family, 'many of my mother's friends are big researchers and scientists, they work a lot, they studied hard in Sri Lanka and then they got scholarships to study in the US, but then they stayed'. Maybe over time, such role models next to her experiences through the science fair projects made her re-envision herself in new ways. It was certainly a network of practices very different from that of Alana, as summarised in Figure 2.

Most striking is the manner Mohini's participation in the science fair and winning it, suddenly made a vast number of powerful science practices accessible to her. She went to a hospital, visiting their scientific laboratories and interviewed some of its scientists. She also got to know the local Cancer Institute. That travel made new spaces of science accessible to her that then also became crucial in her self-authoring as a researcher one day, 'maybe in medicine, but then I like physics a lot too, it really interests me'.

What made her win the gold medal at the local, provincial and trans-Canada competitions was the unique manner whereby she approached the study of antimatter and cancer treatment: 'I went beyond just documenting how it works, I went further, I made my proper conclusions, finding things, comparing things, and all that'. She already had ideas on how to expand her study into an experiment, her objective for next year. She enjoyed the possibility to 'get into science, work on it at a less superficial level' than was the case in articles for the newsletter. Note how Mohini's positioning in science mediated her forms of participation in the programme, making her seek out different activities than Alana. Yet, that position

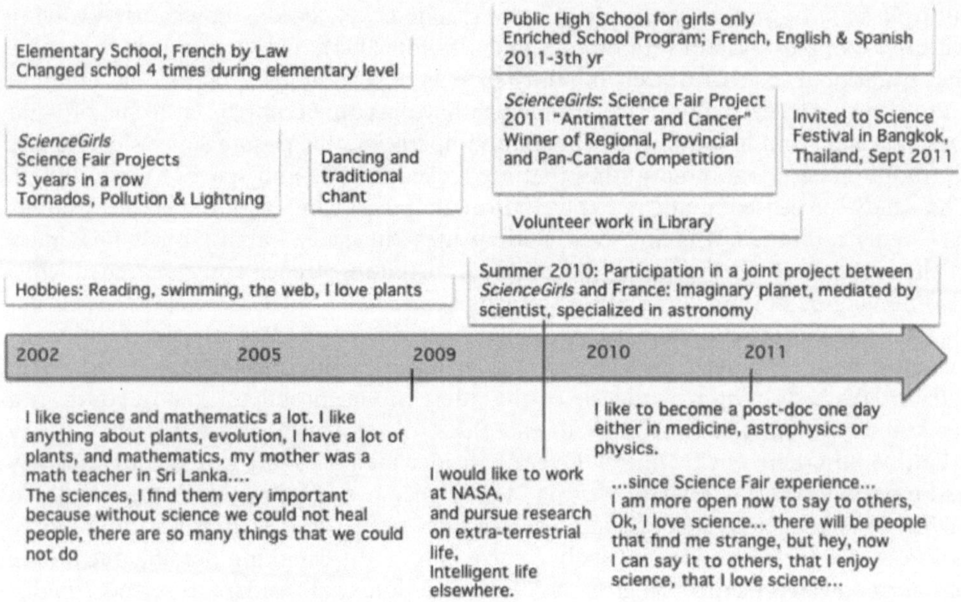

Figure 2. Mohini's repertoires of cultural practices and trajectories of identification.

and navigating inside the programme also became a means into other spaces of science beyond the programme that remained inaccessible to Alana. Mohini's trajectory of self-identification, as apparent in the introductory paragraph and summarised here did not change much over time.

The story I weave together between the two girls and myself is mutually constructed, yet this story is partial and needs to be recognised as such. Through a focus on trajectories of identification in and beyond science, it became clear just how complex the relationship is between learning and identity work. As Nasir and Cooks (2009) contend, I see learning and identity as distinct processes, yet as informing one another. Through engagement in science in the programme the two girls had access to different kinds of identities in science. Yet, the making of identities was not only located in that time-space dimension nor can it be reduced to the availability of identity positions the girls aspired to. Their identity work was also marked by the kind of cultural and social capital they had access to in the past and at the moment and their learning ecologies or network of practices that constituted their lives. Agency was tangled up with positioning too. Since being an insider to science did not mean the same thing to both girls, it made them seek out the spaces in which they participated in different ways. The two cases underline the importance of much more nuanced views of trajectories into science. There are many pathways into science and ways to be inside science, yet some have more power and hence, are recognised publicly (case of Mohini) while others are marginalised (case of Alana).

Discussion: multi-sited ethnography, science literacy and equity – a possibility

The stories about science literacy development and identity work in science in the context of a multi-sited ethnography point to ways science is best understood as a cultural activity and as socio-historically situated, yet also as constantly evolving within a complex system (Hammond and Brandt 2004). This paper made the case that traditional units or objects of study such as science literacy, culture, community and science learners and identity in science have become complex in terms of what they represent and how they are understood operationally, posing serious challenges to traditional ethnographic studies that are locked in time and space (Marcus 2009). The study presented made us travel through places and times. Yet, as Marcus repeatedly cautioned, it is not about adding sites but instead about a new imaginary of the research scene that calls for new roles of the researcher and researched while also moving us beyond description towards action.

The collaborative imaginary described here made possible an examination of meaning making of self and science 'in the wild' (Hutchins 1995, referenced in Barron 2006), offering an analysis of the interactional practices among youth in a space they owned. The possibility to enact the role as youth assistant drew some in (Alana) and others out (Mohini). Overall, insider identities in the newsletter activity had more to do with a love for writing than science per se, as Mohini's trajectory out of the newsletter activity back into science fairs suggests as well. Yet, did the girls' positioning of selves as journalists become a commodity, making possible the hiding of their contested identity as girls of colour and potential insiders to science? Did it make engagement with science possible in ways they judged safe or was it really due to their value of science and interest in putting science to use, rather than engage in science as an end in itself, as maybe Mohini did, who moved out of the newsletter activity into the science fair in ScienceGirls?

These questions and concerns point to the current challenges brought about by the post-industrial global economy that intensified the schooling's role in the production of inequality, acting as an ever more tight and effective sorting machine, which led to an increase in studies of spaces that youth carve out for themselves (Vadeboncoeur 2009), that are less alienating than schools, and that make engagement in science and significant learning possible. ScienceGirls are 'counter publics' or 'safe spaces' that offer opportunities to youth to put their imaginative resources to work. They are also a crucial 'third area' of youth and literacy development, besides school and family (Halpern 2003; Heath and McLaughlin 1993). Yet, at what cost?

While single space-based research of informal learning is important, it rarely questions the role of such learning in a larger network of practices and within a web of practices differentiated by power. Some studies on informal learning have also endorsed a spatial dichotomy between school and CBOs. Yet, as suggested by Weis and Dimitriadis (2008, 2299), those studies together have led to 'an ironic scholarly moment of valorizing productions of out-of-school cultural vibrancy without ever asking how such vibrancy articulates with, or perhaps even smacks in the face of, an increasingly tight "sorting machine," broadly defined', leaving us with no 'sense of how these sites (formal and other) are enmeshed in complex webs of relationships for their participants'. The same can be said about science literacy development. There have been few attempts in trying to understand how engagement in science in school, the family, in CBOs and outside constitute youths' science literacy development and identity work in science over time. Similarly, the manner youths' biographies of engagement with science are culturally and politically marked in an era of globalisation needs unpacking. As is, science education notably fails to engage women and youth from high-poverty and non-dominant communities. They continue to be underrepresented in certain science professions and areas in science in higher education (Bell et al. 2009; Calabrese Barton 2007). Girls in underserved communities also have less access to and participate in smaller numbers in well-structured leisure-time activities that entail science (Fadigan and Hammrich 2004). Together, these arguments make the case for intervention at many levels of a complex political system to bring outsiders in.

Furthermore, few studies have examined how the accrual of such experiences translates into the capacity to use scientific knowledge for personal and social purposes throughout life. While research suggests that informal learning environments are particularly important to marginalised youth in offering them – often for the first time – to engage in and with science in to them personally meaningful ways, such engagement in science and identity work does not translate into agency, empowerment, social just science or economic mobility for most marginalised youth (Calabrese Barton 2007; Rahm 2010). It seems imperative to more effectively understand the manner social class, ethnicity, community, culture and history intersect and constitute learning and identity work in science yet also mobility within and beyond the world of science (Bell et al. 2009). To do so, multi-sited ethnography and certainly longitudinal studies are needed that make possible travel with youth across space and over extended periods of time. Research is needed that attempts to unpack connections and barriers in learning across practices, shedding light on the hybrid nature of youths' engagement in science in a world marked by globalisation. It calls for a focus on how time-space configurations socially construct learning and who can become an insider to science. The study underlines clearly the complex time-space configuration of science literacy development. As it is, some participants in ScienceGirls received public recognition for their work. Yet, for most (like Alana), participation did not translate into more mobility in spaces of power that counted. Alana's positioning as an insider in a science that leads to action next to her active contributions to such forms of engaging with science remained unnoticed in school too, where she started to struggle academically in mathematics, a topic she perceived as key for getting into what she called the hard sciences. Yet, based on our data, it is impossible to assess to what extent her positioning in science was tied to a lack of interest in the hard sciences as defined by her, or tied to her perceived inaccessibility of science.

As discussed, multi-sited ethnography also challenges the anthropologist-other binary, conceiving of it as a collective, and hence, hints at a participatory research imaginary (Cammarota and Fine 2008; Delgado 2006). That is, participatory youth research may entail a natural extension of a multi-sited research imaginary and could offer new and important insights into issues of equity in science education and answers to the many questions raised by this research imaginary, some of which I noted above. It could help us understand the extent to which the level of the girls' social position as first- and second-generation immigrants constitutes their relation with science. Was the programme marked by 'transgression and reproduction' or was it instead a safe space to explore and develop identities in science that would potentially carry over into other practices (Holland and Lave 2001, 19)?

Collaborative participatory action research could also help us move towards a focus on promising practices of social just science and thereby push the dialogue further, beyond accessibility concerns in science. In such a research imaginary, youth would no longer be objectified. Instead, they would actively contribute to the research imaginary and be given a voice in the constructions that emerge about who they are, their community, and what science literacy may entail as they live it as first- or second-generation immigrants. Youth participatory action research helps deconstruct 'who can "do" research, what constitutes research and who benefits' (Cammarota and Fine 2008, 8). Through participatory research, youth not only learn about the ways to conduct research by conducting it, but also engage in finding solutions to problems and thereby collaborate in significant ways in putting research findings to use. Such research endeavours would give voice to the ones still marginalised today in conversations about equity in science education and what such may imply and for whom, that has started to emerge in current research in science education (see for instance, St. Louis and Calabrese Barton 2002).

As argued throughout, we can no longer leave unquestioned the complex time-space configuration of science literacy development, its inherent trajectories of identification and learning, and its role in the reproduction of race and class structures in science education. Yet, we also need to further expand opportunities for meaningful engagement with science and critical reflection on how to then transform such ways of knowing, doing and being into capital for navigations that are empowering for diverse youth and open up new possibilities for being in science.

Acknowledgements

I thank the youth and research team that contributed to the imaginary, especially Audrey Lachaîne, Allison Gonsalves, Anne Gorry, Itzel Vazquez and Fasal Kanouté. The research was supported financially by the Social Sciences and Humanities Research Council of Canada.

References

Aikenhead, G.S. 2006. *Science education for everyday life: Evidence-based practice.* New York: Teachers College Press.

Barron, B. 2006. Interest and self-sustained learning as catalysts of development: A learning ecology perspective. *Human Development* 49: 193–224.

Basu, S.J., A. Calabrese Barton, and E. Tan, eds. 2010. *Democratic science teaching: Building the expertise to empower low-income minority youth in science.* Rotterdam: Sense.

Bell, P., B. Lewenstein, A.W. Shouse, and M.A. Feder, eds. 2009. *Learning science in informal environments: People, places, and pursuits.* Washington, DC: The National Academies Press.

Calabrese Barton, A. 2007. Science learning in urban settings. In *Handbook of research on science education*, ed. S. K. Abell and N. G. Lederman, 319–43. Mahwah, NJ: Lawrence Erlbaum.

Cammarota, J., and M. Fine. 2008. Youth participatory action research: A pedagogy for transformational resistance. In *Revolutionizing education: Youth participatory action research in motion*, ed. J. Cammarota and M. Fine, 1–11. New York, NY: Routledge.

Delgado, M. 2006. *Designs and methods for youth-led research.* Thousand Oaks: Sage.

Deutsch, N.L. 2008. *Pride in the projects: Teens building identities in urban contexts.* New York, NY: New York University Press.

Dolby, N., and F. Rizvi. 2008. *Youth moves: Identities and education in global perspective.* New York, NY: Routledge.

Fadigan, K.A., and P.L. Hammrich. 2004. A longitudinal study of the educational and career trajectories of female participants of an urban informal science education program. *Journal of Research in Science Teaching* 41, no. 8: 835–60.

Gutiérrez, K., and C.D. Lee. 2009. Robust informal learning environments for youth from nondominant groups. In *Handbook of research on literacy and diversity*, ed. L.M. Morrow, R. Rueda, and D. Lapp, 216–32. New York: The Guilford Press.

Gutiérrez, K., and B. Rogoff. 2003. Cultural ways of learning: Individual traits or repertoires of practice. *Educational Researcher* 32: 19–25.

Halpern, R. 2003. *Making play work: The promise of after-school programs for low-income children*. New York: Teachers College Press.

Hammond, L., and C.B. Brandt. 2004. Science and cultural process: Defining an anthropological approach to science education. *Studies in Science Education* 40: 1–47.

Handa, A. 2003. *Of silk saris and mini-skirts: South-Asian girls walk the tight-rope of culture*. Toronto: Women's Press.

Harvey, D. 1990. Between space and time: Reflections on the geographical imagination. *Annals of the Association of American Geographers* 80, no. 3: 418–34.

Heath, S.B., and M.W. McLaughlin, eds. 1993. *Identity and inner-city youth*. New York, NY: Teachers College Press.

Holland, D., W. Lachicotte, D. Skinner, and C. Cain. 1998. *Identity and agency in cultural worlds*. Cambridge, MA: Harvard University Press.

Holland, D., and J. Lave, eds. 2001. *History in person*. Sante Fe, NM: School of American Research Press.

Holland, D., and K. Leander. 2004. Ethnographic studies of positioning and subjectivity: An Introduction. *Ethos* 32: 127–39.

Hughes, G. 2001. Exploring the availability of student scientist identities within curriculum discourse: An anti-essentialist approach to gender-inclusive science. *Gender and Education* 13: 275–90.

Kincheloe, J., and K. Berry. 2004. *Rigour and complexity in educational research: Reconceptualizing the bricolage*. New York, NY: Open University Press.

Knobloch, U. 2009. Meta-analysis of gender and science research: Country report Switzerland. Report, Seventh Framework Program. Université of Fribourg, Switzerland. http://www.genderandscience.org (accessed October 14, 2011).

Leander, K.M., N.C. Phillips, and K.H. Taylor. 2010. The changing social spaces of learning: Mapping new mobilities. *Review of Research in Education* 34: 329–94.

Lemke, J.L. 2000. Across scales of time: Artifacts, activities, and meanings in ecosocial systems. *Mind, Culture, and Activity* 7, no. 4: 273–90.

Marcus, G. 1998. *Ethnography through thick and thin*. Princeton, NJ: Princeton.

Marcus, G.E. 2009. Multi-sited ethnography: Notes and Queries. In *Multi-sited ethnography: Theory, praxis, and locality in contemporary research*, ed. M.-A. Falzon, 181–96. Burlington, VA: Ashgate.

Nasir, N., and J. Cooks. 2009. Becoming a hurdler: How learning settings afford identities. *Anthropology & Education Quarterly* 40: 41–61.

Nespor, J. 1994. *Knowledge in motion: Space, time and curriculum in undergraduate physics and management*. Washington, DC: The Falmer Press.

Nespor, J. 2004. Educational scale-making. Pedagogy. *Culture and Society* 3: 309–25.

Phipps, A. 2007. Re-inscribing gender binaries: Deconstructing the dominant discourse around women's equality in science, engineering, and technology. *The Sociological Review* 55, no. 4: 768–87.

Pierides, D. 2010. Multi-sited ethnography and the field of educational research. *Critical Studies in Education* 51, no. 2: 179–95.

Rahm, J. 2010. *Science in the making at the margin: A multisited ethnography of learning and becoming in an afterschool program, a garden, and an Upward Bound Program*. Rotterdam: Sense Publisher.

Schiebinger, L. 1999. *Has feminism changed science?* Cambridge, MA: Harvard University Press.

Seiler, G., and A. Gonsalves. 2010. Student-powered science: Science education for and by African American students. *Equity & Excellence in Education* 43: 88–104.

Soja, E.W. 1996. *Thirdspace: Journeys to Los Angeles and other real-and imagined places*. Malden, MA: Blackwell.

St. Louis, K., and A. Calabrese Barton. 2002. Tales from the science education crypt: A critical reflection of positionality, subjectivity, and reflexivity in research. *Forum Qualitative Sozialforschung/Forum: Qualitative Social Research*, 3, no. 3, Art. 19. http://nbn-resolving.de/urn:nbn:de:0114-fqs0203196.

Urrieta, L. Jr. 2007. Figured worlds and education: An introduction to the special issue. *The Urban Review* 39, no. 2: 107–16.

Vadeboncoeur, J.A. 2009. Spaces of difference: The contradictions of alternative educational programs. *Educational Studies* 45: 280–99.

Weis, L., and G. Dimitriadis. 2008. Dueling banjos: Shifting economic and cultural contexts in the lives of youth. *Teachers College Record* 110, no. 10: 2290–316.

The role of borders in environmental education: positioning, power and marginality

Carrie T. Tzou[a] and Philip Bell[b]

[a]*Education Program, University of Washington Bothell, Bothell, WA, USA;* [b]*Learning Sciences & Human Development, College of Education, University of Washington Seattle, WA, USA*

In this study, we argue that in order to understand the underlying causes of inequities in education, we need to look outside specific educational settings to the larger social, historical and political structures behind those inequities. Therefore, in this work we take a social justice perspective to look at the relationship between environmental science education, dominant environmental narratives and how places and youth are constructed and positioned through environmental education. To this end, we examine how *borders* – and their corresponding margins – construct place in environmental education and to discuss implications of this construction of place for issues of environmental justice and equity. Borders both demarcate categories and serve to reify existing social and political relationships. Therefore, we ask the following questions: (1) What is the role of borders in constructing place in environmental education? (2) How do youth position themselves with respect to these borders? and (3) What are implications for the design of environmental education learning environments that work towards goals of social and environmental justice? We found that prevailing environmental education narratives construct borders using what we call discourses of fear and privilege, and that these boundaries position communities and youth in ways that reify existing social and political power relationships. We also found that youth resist these constructions both through counter-script and performance of identity.

Introduction

It is well-documented that exposure to environmental hazards are experienced differentially by communities along race and class lines (Bullard 1990) and that there are corresponding inequities in the rates at which communities experience health problems such as asthma due to exposure to environmental pollution (MacLaren and Houston 2004). Furman and Gruenewald (2004) argue that 'environmental crises are inseparable from social crises' (48), yet, traditional environmental education has yet to take up issues of environmental justice and multiple cultural perspectives, thus serving to reinforce dominant societal values and marginalise those groups of youth who are most impacted by the most pressing modern environmental issues (Grass and Agyeman 2002). In this paper, we use the construct of *borders* to study both how youth resist dominant environmental narratives that serve to disconnect them from the places in which they live, and how they assert identities as separate from the activity of environmental education.

We take a social justice perspective (Calabrese Barton 2003; Giroux 1992) to look at the relationship between environmental science education, dominant environmental narratives, and how places and youth are constructed and positioned through environmental education. To this end, we examine how borders construct both the centres and margins of place in environmental education and to discuss implications of this construction of place for issues of environmental justice and equity. Borders are an interesting construct because of the ways in which they demarcate categories and serve to reify existing social and political relationships (Jones 2009). We ask the following questions:

(1) What is the role of borders in constructing place in environmental education?
(2) How do youth position themselves with respect to these borders?
(3) What are implications for the design of environmental education learning environments that work towards goals of social and environmental justice?

Theoretical perspectives

In this paper, we explore three main theoretical constructs: *place, borders and identity.* We use literature from anthropology, sociocultural studies in education and cultural geography to define *place.* Specifically, one perspective in anthropological literature argues that places are 'politicised, culturally relative, historically specific' and 'socially constructed' (Rodman 1992) and that places are perceived by us through our own cultural lenses and value systems (Casey 1996). Places teach us how to be in the world and our place in the world (Gruenewald 2003). From a sociocultural perspective, places are 'a complicated, ecological system that includes physical, biological, social, cultural, and political factors with the history and psychological state of the persons who share the location' (Lim and Calabrese Barton 2006). Places are therefore a complex interplay of physical and cultural factors that are perceived by us based on our experiences and identities.

In contrast to being merely a 'context *for* society – its container' (Soja 1980, 210), place simultaneously is structured by and structures human activity. Therefore, there is no social activity outside of places, and place has no meaning outside of intentional social practice. In the places where environmental education occurs, those places embody environmental narratives and norms that 'permit fresh actions to occur, while suggesting others and prohibiting yet others' (Lefebvre 1991, 73). From this perspective, actions within environmental learning settings are not neutral or random; they are, instead, situated within historical and spatial contexts that invite or prohibit opportunities for learning.

Borders, identity and marginality

We take *borders* to be distinctions whose main effect is to demarcate one group from another – to create categories. Newman (2006) writes that, 'borders create (or reflect) difference and constitute [a] separation line' (148). Much attention has been paid to borders in fields such as cultural and political geography (Newman 2006; Newman and Paasi 1998), especially as they pertain to separation between political states and resulting identities that develop within such boundaries. However, in this paper we merge the notion of borders from a cultural geography sense with ideas from sociocultural and social justice literature in education (Calabrese Barton 2003; Gruenewald 2003; Gutiérrez, Rymes, and Larson 1995) to explore the ways borders create what Newman and Paasi (1998) call *sociospatial identities* that construct place and position youth in certain ways. Specifically, we will look at how environmental narratives demarcate acceptable and safe environmental actions from those that are dangerous and toxic and how this demarcation positions youth and their communities in disempowering ways.

Borders as a construct are interesting to us not only because of their role in environmental education but also because of what Jones (2009) calls the *paradox of categories*. Borders are essential to making sense of the world – the world would be endlessly complex and meaningless without them (Newman and Paasi 1998). However, in their use, ethnicities, cultures, races, nations simultaneously become simplified and homogenised, 'organising diversity into a few distinct containers' (Jones 2009, 178).

Therefore, the paradox, as Jones argues, is that while borders help make sense of the world, in their use the borders act as fixed and closed. Societies exist on the basis of placing people and things into categories (Jones 2009). By using categories that are created by borders to make sense of the world, we find it hard to think outside of those categories. We can then ask, from a critical perspective, *who* has the power to make such categories? Whose realities and interests do the borders reflect? Considerable power comes from the having the right to impose categories on the world 'to make and unmake groups' (Bourdieu 1991, 221, as quoted in Jones, 2009), 'to establish the boundaries between what is on the inside and the outside' (Jones 2009, 177). Therefore, there is no discussion of borders without a corresponding discussion of power (Newman and Paasi 1998).

With borders comes an examination of action that exists at the *margins* of narratives and the categories they create. bell hooks (1990) makes a distinction between being positioned at the margins, or being marginalised to a place of deprivation, and actively *choosing* the margins as a 'site of radical possibility, a space of resistance' (149). Understanding the ways in which some narratives of environmental education form certain boundaries around activities allows us to see how youths' places are either positioned at the margins or at the centre of that activity. We can also understand, through studying situated action, how youth might resist these environmental narratives by positioning *themselves* at the margins to produce what hooks calls a 'counter-hegemonic discourse' (149).

Individuals can also be involved in border creation – demarcating who they are versus who they are not through the performance of their identities in key places. Brickhouse and Potter (2001) argue that 'identities are maintained in performances' (966). Therefore, we also look at these performances as statements of one's identity that define who that individual is or is not. Veninga (2009) calls this 'embodied practices' (116) – the ways in which individuals *do* or *perform* who they are. In this study, we look at a mini case study of one youth, Miguel, and how the performance of his identity was closely related to his participation in the environmental education activities.

Finally, much can be learned from studying how and when the categories of differences are created (Newman 2006, 148). In this process, we may observe characteristics of borders such as how open or closed they are, who is on either side of the border, and who has the power to form the distinction. In this study, we will use both the construct of borders and place to study the *process* by which borders are constructed in environmental education to youth and their communities in a broader environmental narrative.

Methods

Ethnography as a counter-hegemonic tool

We use ethnographic methods in this study (participant observations, interviews, field noting) to deeply understand *place*, or context, as it relates to environmental science learning and youths' behaviours across settings. Fundamentally, we are interested in the ways in which youth and their places are positioned within the rhetoric of environmental science education – whether they are marginalised or made part of the dominant discourse – and how they perform identity in those moments.

Wolcott (1999) argues that ethnography can be a way of *seeing* (as opposed to just *looking)* that involves describing what people do, the meaning they attach to what they do, out of which come some conclusions about cultural process. In this study, we view ethnography also as a counter-hegemonic tool, a way of taking 'norms and expectations as starting points for questioning them, in other words, tak[ing] them as problems rather than as facts]' (Blommaert and Jie 2010, 11). Specifically, we take a particular model of environmental science learning, one that is very didactic and filled with strong environmental narratives of what learners ought to believe about hazards in their community, and question the ways in which it constructs particular notions of place and learners' positions in those places. In this way, ethnography becomes a way of attempting to represent multiple 'realities' other than those represented in some environmental narratives (Goodall 2000; Tzou, Scalone, and Bell 2010) – namely, what Goodall (2000) calls the 'grand narrative of Western, mostly white, mostly male, science' (12). Therefore, ethnography becomes a way of seeing certain narratives of environmental education from the youths' point of view, how it does or does not fit with their lived realities, and a way of consequently theorising about place and borders in relation to those narratives.

Research site

This study took place in an environmental justice programme for high school students. The Riverside Park Community Centre is in a large urban city in the Pacific Northwest. The Riverside Park community itself lines the banks of an industrialised river that was designated by the Environmental Protection Agency as a Superfund site, which means that it is one of the most toxic waste sites in the country. However, it is part of a larger watershed system that drains water into a large estuary that then connects to the Pacific Ocean. Salmon, fishing, and shellfish are part of the region's cultural and economic health, and the superfund site has become a nexus in the community for discussing these issues as well as the cleanup of the river itself.

Riverside Park is one of the few communities left in the city that is zoned for both commercial and residential use; therefore, commercial companies can be physically located next to private residences. The rate of hospitalisation due to asthma in Riverside Park is over two and a half times the county rate (County Department of Health 1997). According to census data from 2000, 37% of the residents identify as

Hispanic/Latino(a), the largest Hispanic/Latino population in the city (City Department of Planning and Development 2006).

The environmental justice youth programme was sponsored by a local, non-profit organisation to educate youth about issues in environmental justice using the Superfund site as a context for learning. The structure of the programme was comprised of 2-hour after-school classes held at the community centre every Wednesday and community service activities every Saturday for 4–6 hours. Each week, the after-school portion of the programme involved a speaker (or sometimes two) to talk about an environmental issue such as environmental justice, salmon habitat, household toxins and gangs. The community service days took included habitat restoration, shovelling compost, and working at the local Green Fest. Table 1 summarises the topics in both the community service days and the Wednesday evening classes.

Data collection and analysis

Wolcott (1999) argues that ethnographic research allows us a way of *looking* that includes both 'experiencing', 'enquiring'. In this study we positioned ourselves as participant-observers, taking hand-written notes during every session. When the youth pulled blackberry bushes, volunteered at the local convention on the environment, or hauled compost, we were participants with them. In the process, we jotted down accounts of the instruction, interactions and stories of the participants as they navigated through these varied settings. Whenever possible, verbatim talk was captured in the field. After each session, these field accounts were written up as ethnographic field notes (Emerson, Fretz, and Shaw 1995) and used as the main data source for this analysis. This position allowed us a way of *experiencing* the events in the programme from the position of the youth in the group. By so doing, we took in information, as Wolcott (1999), says, 'from all of the senses', from listening to comments and conversations from the youth, to smelling the compost and feeling sore muscles as we shovelled it, to the emotional exhaustion from being bombarded with heavy environmental messages during classes.

Table 1. List of settings in which youth were observed during the environmental justice programme and class topics for Wednesday evening classes. The topics in bold were the topics that were presented in a PowerPoint/lecture format.

Community service settings	Wednesday evening class topics
• Polluted river site/tour of community • Pulling blackberry bushes at local creeks • Local university • Three-day sailboat trip • Earth Day volunteer event at local watershed • Local 'green' festival • Riverside Park Community Centre • Local aquarium • Wastewater treatment plant • Local farm/Riverside Park community for graffiti cleanup	• History of the community & the river • Salmon habitat • Environmental justice • Toxins in homes (RQ episode) • Getting ready for the sailboat trip • Gangs • Toxic household cleaners • Mexican history and culture

Our *enquiring* came in the form of both semi-structured interviews and what Wolcott (1999) calls 'casual conversation'. Especially during the community service sessions, these casual conversations allowed us to ask questions about previous experiences, how the experiences in class related to experiences at school, and the youths' perspectives on events happening in the moment. In addition, more formal, semi-structured interviews (Spradley 1979) were conducted with a focus group of participants at the end of the study. The purpose of these interviews was to ask youth about key incidents in the programme and to member check about our interpretations of these events. For example, Miguel, a youth highlighted later in the paper, was a key informant both in casual conversations and in the semi-structured interviews, as he was outspoken about his opinions about events in the programme (e.g., being smarter than the other people in the class, why there were no people of certain ethnicities doing community service, etc.) and we later asked him about those opinions in the semi-structured interviews.

Field notes were analysed for evidence of categories being attributed to objects, places, or people in order to study how boundaries were made to delineate categories from each other. We used an open coding method (Strauss and Corbin 1998) to look for patterns in the ways in which the talk created borders or distinctions. We also analysed field notes for moments in which the youth themselves were asserting certain aspects of their identities. Together, these moments added up to profiles of identities in which youth were forming personal boundaries between themselves and the world.

Findings

We observed borders-in-the-making along several dimensions, all of which have implications for design of environmental education for social justice. The findings that we will present in the following sections are:

(1) Borders were constructed by the environmental educators using what we call rhetoric of fear and privilege. These borders were used to categorise products and action that were safe, healthy and acceptable and those that were unhealthy for the environment and personal health.
(2) We found evidence of youth resisting these borders-in-the-making through what Gutiérrez, Rymes, and Larson (1995) call *counter-script*.
(3) Finally, we will present a mini case study of Miguel, who we found resisting from the margins in the performance of his identity as a Mexican. This identity construction served to both distinguish and separate him from the activity systems of the programme.

'Pthalates are the everywhere chemical': rhetoric of fear and privilege

The notion of borders was originally interesting to us because of the barrage of messages the youth received around the hazards present in their environment – from the pollution of the river to the destruction of salmon habitats, to the danger of everyday personal care and cleaning products. Sobel (1999) writes that:

> In our zest for making them aware of and responsible for the world's problems, we cut our children off from their roots... My fear is that our environmentally correct curriculum will end up distancing children from rather than connecting them with, the natural world. (1–2)

What is missing from some environmental education is a narrative of connectedness to place to support youth in forming positive relationships with the places in which they live and thus to inspire them to want to conserve those places. Giroux (1992) argues that in education, over-emphasis on what he calls the 'language of critique' results in 'exhaustion and despair' (78). It is therefore in the combination of critique and possibility that environmental education can move beyond 'despair' of the world's environmental problems towards positive action and empowering youths towards enacting positive change.

In five of the seven Wednesday night sessions of the programme, speakers followed a PowerPoint/lecture format (see Table 1). We will elaborate on one of these instances in order to do a deeper analysis of how one border in particular was formed. This was representative of the structure of the talks. The following is a sample from a presentation made by to the students in a Wednesday night session of the programme. The speaker (RQ) came from a local organisation against household toxins and made her presentation through PowerPoint slides.

1 RQ [speaker]: chemical ingredients in kids' products are not always chemically bound to products – they enter kids' bodies through mouths, skin contact, off-gassing into the air. [talks about lead and brain development] ADD: could some of that be due to tiny amounts of lead as a kid? [holds up little slinkies in a plastic bag]: little plastic slinkies bought at Target, it turns out they have 1700 ppm of lead.	
	2 Sarena: Can I have one?
3 RQ: No, I don't want them out there.	
	4 Sarena: I used to play with these!
5 RQ: Phthalates are the everywhere chemical. Potential side effects are asthma, damage to lungs or kidneys, and impacts on the male reproductive system. Women getting exposed might have male babies with birth defects. If you pick up a bottle of body wash (picks up a bottle of body wash)	
	6 Rosanna: Oh, I use that!
7 RQ looks at the chemicals listed on the bottle	
	8 Sarena: Can I get that? I ran out!
9 RQ: Well I'll let you decide for yourself if you still want it. [she reads down the list of chemicals and passes the bottle around]	

10 RQ: Examples of chemicals of concern in cosmetics/personal care products are carcinogens, endocrine disruptors/ reproductive toxins, neurotoxins.	
	11 Sarena: I had the perfume. I had it last year.
12 RQ: Phthalates are found in 75% of products: perfumes, deodorants, hair gels, body lotions. People can't even look at the label and be sure the product doesn't have phthalates in it.	
13 RQ: I worry about people that work in the nail salons because they get expose all day long.	
14 You live so close to the River and the River connects to the Puget Sound. So all these products are going down the drain. They don't get broken down in wastewater treatment. (She mentions the wastewater treatment some things pass through unchanged.) Sometimes it's not good for the health of fish that these are out there. Fish biologists worry about that some populations of fish don't have the right proportions of males and females. Some of these fish are getting feminised early in development.	
	15 Sarena starts giggling
16 In the River cleanup, phthalates are what we are trying to clean up. Even thought we try to clean it up, it gets re-contaminated because phthalates are used in everything. Isn't it funny to think about that the things you use in your bathroom can affect the fish populations in the Puget Sound?...	
17 Jose asks about hair gel.	
18 RQ: the problem is that almost every product for hair has some fragrance in it and that almost always has phthalates.	
	19 Miguel: you're not dead so it's ok.
	20 Gabriela [instructor]: We always say buy deodorant – not aluminum. In Mexico, I see people using lemon (makes a gesture like she's using a lemon as deodorant]. They don't have the money.
	21 Miguel: That's old school.
	22 Juanita: pass me an orange. [ss laugh]
23 RQ: we're all being marketed to – there's a lot of commercials telling me I have to use these products. You have to ask what do I want to do?	
	24 Miguel: I want to smell good.

In this example, we can begin to see how RQ's narrative of dangerous household toxins constructs a border between safe and unsafe products – both for personal health and for the environment. One of the main strategies RQ uses in this case is what we call a *rhetoric of fear*. This includes at least three elements: (1) using scientific evidence and jargon to communicate the seriousness and the omnipresence of the problem, regardless of how or if the students understand it ('1700 ppm of lead', 'carcinogens', 'endocrine disruptors', 'neurotoxins'), (2) communicating the inevitability of coming into contact with these dangerous chemicals ('Phthalates are the everywhere chemical'); and (3) the omnipresence of unsafe products: slinkies from Target [where, it was indicated to us by one of the instructors, almost all of the students shop], body wash, lotion that are available in the local stores.

We argue that at the same time that these strategies serve to categorise safe (chemical free) and unsafe (phthalate-ridden) products, they simultaneously construct and position *place* in particular ways. By invoking the River starting in line 14 and saying 'You live so close to the River and the River connects to the Puget Sound. So all these products are going down the drain. They don't get broken down in wastewater treatment'. The implication is that if they do not get broken down, they end up in the River again and close to their homes. RQ thus correlates the danger of the chemicals with the environmental danger of phthalates in their local community. She says, 'Isn't it funny to think about that the things you use in your bathroom can affect the fish populations in the Puget Sound?', thus simultaneously making light of the seriousness of the problem ('Isn't it funny?') and communicating the nearness and seriousness of the impacts of the problem, from their bathrooms to the Puget Sound.

Positioning theory (Harré and Moghaddam 2003) helps us make sense of this episode in terms the range of activities that the presenter made available in her speech act of conveying information to the youth. They argue that the social force of an action and the position of the actors and interactors mutually determine one another (17). In addition, because the products that RQ uses to demonstrate 'unsafe' products are those that are readily accessible to students in their homes and grocery stores, this indirectly positions the community as one that is unsafe. If the products that students use in their homes are filled with these fearful chemicals, how, then, should students view the safety of their homes and their communities? Who has the power to construct their homes and places in this way?

Another strategy at play in this example is what we call the *discourse of privilege*. While it may be true that plastic toys from Target contain lead and other chemicals – and that these chemicals are potentially hazardous to our health, alternatives to these products are not always readily available. The *discourse of privilege* implies that we should heed certain environmental narratives and avoid these products and that we, in fact, have the choice and the ability to do so. This is not always true. In the Riverside Park neighbourhood, in fact, the major retailers are Target and small grocery stores. Lotions and body washes that are 'chemical-free' are largely more expensive and unavailable in the small grocery/convenient stores in Riverside Park. Low-income communities often pay more for inferior – or, in this case, 'unsafe' – products in their grocery stores (Komunyakaa 2002). Therefore, the rhetoric of privilege makes invisible and takes for granted the privileged stance from which environmental messages arise and thus marginalises the lived realities of the youth living in this community.

The combination of these two rhetorical strategies marginalises the readily available, 'unsafe' options and places to live, thereby rendering youth with few options to affect change in their own situations as they are surrounded by fearful chemicals that they cannot escape ('People can't even look at the label and be sure the product doesn't have phthalates in it'.) and they may not have the resources in their communities to buy alternatives. We also find it interesting that the courses of 'action' presented to students by RQ are limited to their 'buying power', what they can do with their money. While this is certainly a concrete action that students can take, we argue that this does nothing to help students question the status quo in terms of which products are available to them compared to products in other communities, for example. This perpetuates what Calabrese Barton (2001) argues is a science education that 'emphasises corporate values at the expense of social justice and human dignity' (847). Bowers (2001) similarly argues that eco-justice education should bring to light the 'hyperconsumerist' strategies promoted by elite groups that end 'undermining the heritage of minority cultures' (77).

Therefore, in trying to understand the social construction of this boundary around safe and unsafe activities we can engage in around household products, we can see an example of the *paradox of categories* (Jones 2009). While the boundary is important in helping to communicate environmental education messages and making distinctions between safe and unsafe products, the boundary itself 'fixes' and positions the community in ways the reify existing social and political boundaries.

Disrupting the boundary: counter-script

In the above example, we see moments of youth resisting RQ's discourses of environmental fear and privilege, what Gutiérrez, Rymes, and Larson (1995) call the teachers' *script*. Gutiérrez, Rymes, and Larson (1995) write, 'In the face of a seemingly incontestable teacher script, students assert forms of local knowledge that are neither recognised nor included in the teacher script' (451). They describe the process of a student-created *counter-script* that represents a challenge to the dominant script of the instructor, sparked by sometimes random associations to the teacher's script, and rarely incorporated into the teachers' script. Therefore, what looks to be, on the surface, random disruption on the part of the students can be understood as a systematic challenge to the authority of the dominant script.

Therefore, while RQ largely holds the power, in the above example, to 'make and unmake the world' (Bourdieu 1991, 221, as quoted in Jones, 2009) in this instance, there are also moments of students disrupting this discourse of fear from the margins of activity. This is illustrated in lines 2, 4, 8 and 11 in which Serena interrupts RQ's narrative to display her apparent fearlessness of the products RQ deems unsafe. By inviting these products ('Can I have one?') and saying that she 'used to play with these' or 'ran out', she signals to RQ that these products could be a part of her everyday life (whether they are or not, we do not know) and that she does not see them as unsafe. From her position of marginality, therefore, she is able to challenge RQ's position as the dominant speaker in the interaction at the same time that she attempts to invalidate her claims of the dangerousness of the products.

Another instance of counter-script comes in lines 15–17. Miguel first says to Jose 'You're not dead so it's ok' – implying that the products must not really be that unsafe if they have not killed him yet. Later, in line 19, he says, 'I want to smell good', meaning that he takes smelling good as a priority to considering how unsafe those products might be. Gabriela then offers an alternative – using lemon as deodorant and Jennifer says, 'Pass me an orange' as a way of building off of Gabriela's example. Neither Miguel's nor Gabriela's comments are taken up into RQ's script.

Gutiérrez, Rymes, and Larson (1995) argue that a 'third space' can be created in learning environments, in which a 'true dialogue' between teacher and students can occur and, thus, 'the possibility of contesting a larger societal, or transcendent, script emerges' (453). Had RQ engaged with the youth in the moments of 'disruption' or counter-script, perhaps a third space could have been created where 'multiple literacies', or multiple ways to engage in environmental narratives could have emerged. Through this dialogue, a 'discourse of possibility' (Giroux, 1992) might have emerged, where youth and instructor begin to contest the larger environmental justice issues such as why this might be important from the youths' perspective, why certain products or stores are only available in certain communities, and what possible actions might be to protest the presence of such chemicals in products.

In design-based curriculum work we have begun with teachers and students in large urban high schools, we have started to explore what this 'discourse of possibility' might look like in a practical setting. In this curriculum, called *My Place in Puget Sound*, we explore pedagogical strategies such as self-documentation (Clark-Ibanez 2004; Tzou, Scalone and Bell 2010) to simultaneously connect to the chemistry of Puget Sound and to youths' cultural values and everyday practices. In this way, environmental education becomes about questioning pervasive environmental narratives in terms of how they relate to one's everyday life, and making decisions that are based at the intersection of scientific evidence, personal health and everyday practices (and constraints on those practices, such as money, geographical location and personal values).

If we refer to Figure A as a way to summarise our findings, then, we argue that RQ's speech activity was framed in a very particular environmental and cultural narrative of fear, human health concerns and toxicity of household products. Because of the sheer dominance of her speech act in this educational setting, she positioned youths' experiences, playfulness and resistance at the margins of activity, leaving the youth to invent ways to create what hooks (1990) calls the 'counter-hegemonic' dialogue, or the counter-script.

Resisting from the margins: Miguel, George Lopez and the performance of cultural capital

At the time of the programme, Miguel was a junior in high school. He played football on his high school team and had aspirations to play for the University of Southern California. He was bilingual in Spanish and English, born in Mexico and came to the US when he was two years old. Miguel had a younger sister who was also a regular participant in the programme, and a younger brother for whom he felt great responsibility. Miguel was one of the first youth to engage with us at the class meetings. He was outspoken during the Wednesday class meetings, often one of the only participants to answer presenters' questions, and disparaged others for their silence. He thought of himself as smart and a hard worker as well as a leader, yet thought that he did not get the recognition he deserved because he was not respectful enough of authority.

One aspect of Miguel's identity that he consistently made salient was being Mexican. While we recognise that 'being Mexican' is a complex identity with many definitions (cf. Anzaldúa 1999), we will refer to the definitions of Mexican-ness that Miguel himself performs, in order to understand how the performance of this aspect of his identity is utilised at key moments of interaction. In fact, we argue that Miguel used this aspect of his identity as a way to re-centre himself at the *margins* of the activities of the group.

One instance when this became especially salient was on a three-day sailboat trip we took towards the end of the programme. This boat-based education programme gave youth opportunities to learn how to handle a sailboat and engage with a scientific question of their own choosing over three days. Representatives from the programme visited youth before the trip to help them craft a scientific question and understand the scientific equipment used to collect data, help youth pick the food for the trip, and orient youth to expectations while on board. Because of a glitch in planning, the youth in our programme had planned their scientific question and chosen the food, but that trip was cancelled and our group was split up into two separate boat trips, each 'hosted' by another group who had extra room on the boat. This meant that the 'host' group of students, rather than our students, planned the science activities and food for the trip.

While there were numerous opportunities on the trip for youth to engage in scientific practices (water testing, looking at plankton under a microscope), Miguel and another boy from our group were often observed laying down on the deck of the boat or falling asleep during activities, physically separated from the science activities of the group, speaking in Spanish to either Gabriela or the other youth from our programme. When I asked Miguel what might have made the trip more exciting for him, he said if they had been able to study the animals living under the sea, or the pollution in the river in Riverside Park. When there was downtime during the trip, the other youth from Riverside Park also physically separated themselves from the activity of the larger group. When we docked our boat and engaged in scientific activities off the boat, they retreated to a nearby shelter and disengaged from the group.

In Miguel's case, this behaviour was very different from most other class sessions and community service sessions, where he was outspoken and was one of the only people to answer speakers' questions. He often commented on how other students in the programme (especially his sister) were 'dumb' because they did not engage with the speakers. Therefore, it was a contrast to see him so quiet and withdrawn while the boat was on the water. However, at the end of the day when we went on land to have the campfire and sleep, he performed his identity as a Mexican to the larger group of youth in very public moments:

> Later, as we are making s'mores, Miguel shares with the group that he loves George Lopez, who says that Mexicans shave their heads bald everyday but when 'one crazy white bitch' (Britney Spears) does it, it is all over the news. Later, the youth all share their 'high tide and low tide' moments, and a girl from the other group says that her low tide was our group because she was not good with strangers, but that was also her high tide because there were Latinos there and at her school there were only like three others. To this, Miguel says, 'you're representing' and puts his fist up. Miguel then shares that he is the only Mexican on the football team.

> At breakfast the next morning, when all of the youth are sitting around the table, Miguel says that George Lopez says that in Los Angeles, they are so lucky because they can get their chicken from El Pollo Loco, but in Mexico, they have to kill it themselves.
>
> As we are finishing up, Miguel asks me if I can swim. He says he does not know how. He says 'Mexican style' is to throw you in the pool and then you learn to swim. He turns to Gabriela and asks, 'Verdad?' [isn't that true?] and she nods and says that is what she did with her son – threw some keys in the pool and then threw him in when he was four years old to get them.

While we do not have secondary data to follow up with Miguel about what he actually meant in each of these moments, we can ask what they are *doing* for Miguel, what role they are playing in positioning him in specific ways in relation to the other individuals on board the boat.

Here, we use hooks's notion of marginality (1990) to understand the ways in which Miguel and the other Riverside youth centre themselves at the margins of the environmental activity on the boat trip. Tatum (1997) argues that practices of self-segregation, or choosing the margins, on the part of Black youth can be seen as acts of 'oppositional identity development' (Thomas 2005), and this 'spatial practice of self-segregation' can be seen as a 'politicised action' (Thomas 2005, 1236) in reaction to racism. On our trip, the Riverside youth had no control over the route the boat took (this was dictated by the design of the scientific questions), the food, or the scientific question being asked. We argue that the self-segregation that we observed on the part of Miguel and the other youth from Riverside Park was not that they were unmotivated or not able to engage in the activities, but rather that they felt disenfranchised from the narrative of the boat trip and were asserting power over the situation through this self-segregating act on the margins. Referring back to Figure 1, we argue that the cognitive ecologies that the Riverside youth brought to the trip were not leveraged productively in that place, thus positioning their experiences and interests at the margins of the 'storyline' and constraining the space of possible actions they could take (Harré and van Langenhove 1999). Thus at the margins, they engaged in the spatialised practice of self-segregation as way to resist and/or simply exist at the margins.

In Miguel's case, we see him engaging in what Holland et al. (1998) refers to as *objectification* of aspects of one's culture. Benson (2003) describes 'psychological boundaries of extension', in which an individual asserts identity in terms of 'what belongs to me' (individual control over ideas, privileges, reputation, etc.) and 'ways in which I belong to others' (group identity and solidarity) (63). Miguel, in the above examples, asserts both what belongs to him (his 'Mexican-ness') and who he belongs to (a larger group of Mexicans, including George Lopez). Being the only Mexican on the football team serves to separate him from the larger group. His solidarity with the other youth in our programme, as they physically and linguistically separated themselves from the other group of youth by speaking Spanish and, at times, sitting off the boat in a nearby shelter, separated him from the formal activity systems of the science instruction. West (1993, cited in Solórzano and Villalpando 1998) argues that these kinds of episodes, where cultural capital is invoked that is on the margins of what society acknowledges as acceptable discourse, is a form of re-centring or re-defining cultural capital (Solórzano and Villalpando 1998).

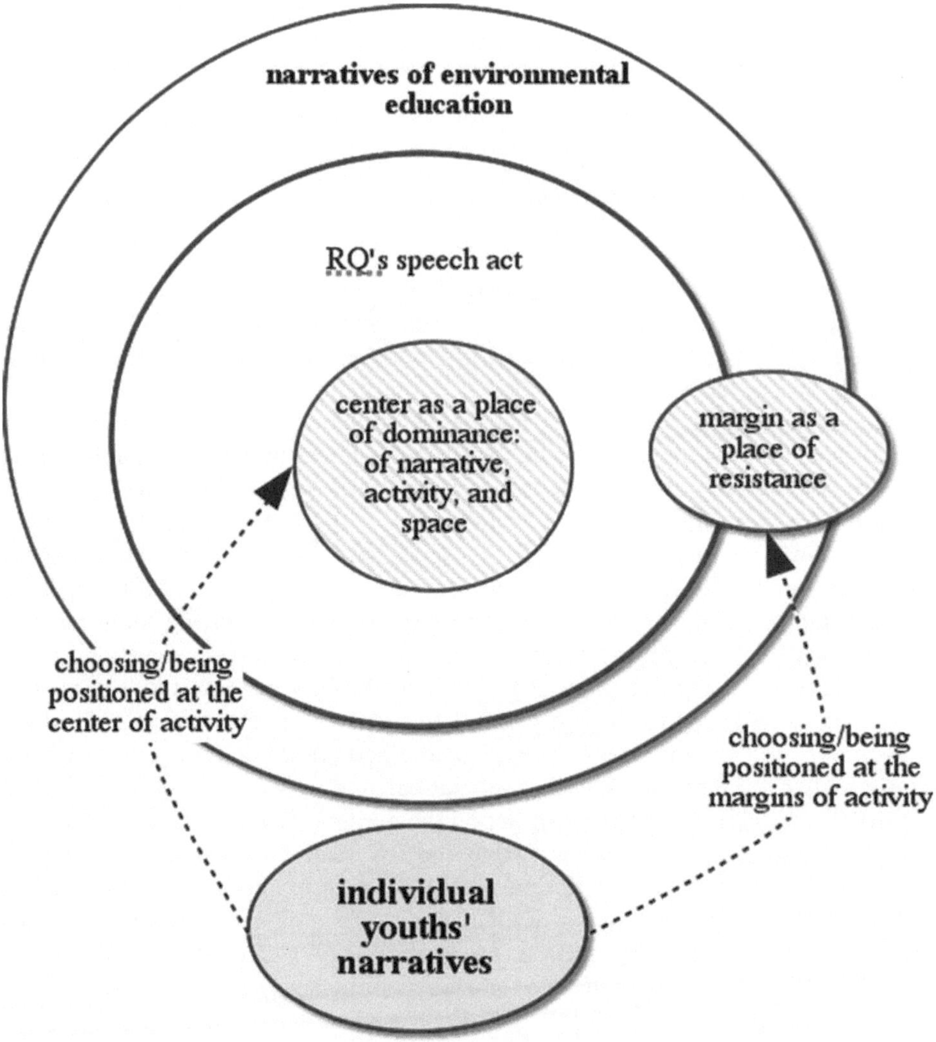

Figure 1. RQ's speech act as situated within the context of a larger environmental narrative. Individual youth come with personal narratives, from which they are either positioned or position themselves at the centre or at the margins of the dominant activity (RQ's speech act).

From a science learning perspective, therefore, we can ask what identities were available (Brickhouse and Potter 2001) to Miguel and the other students from Riverside to assert on the boat trip? What was it about the structure of that place, the borders inscribed by the activity systems and the margins of those systems that drove the emergence of these marginalised actions? We assert that this racialised identity performance stemmed at least partially from the instruction aboard the ship that never took advantage of local knowledge or previous experiences that the Riverside youth had. This brings up questions for us of how learning opportunities can be designed to leverage the identities and everyday knowledge that youth bring with them, as part of their individual ecologies, to the activities of environmental science instruction so that they see those ecologies as part of the centre of the activity? At the same time, however, we realise that this is in tension with the potentially productive ways in which resistance from the margins might fundamentally change scientific and environmental science education discourse if space were opened up to consider it.

In the last day of the programme we were engaged in habitat restoration on a cold, rainy morning. Habitat restoration mainly involved removal of invasive blackberry plants in an area just outside the grounds of an elementary school close to the Riverside Park community. The students were going to do a final presentation on what they learned during the last half of the morning, directly following the habitat restoration. As we were getting our instructions for the day, Miguel again makes reference to his Mexican identity:

> In talking about the final presentation, Miguel says 'Mexicans don't care about the environment. They won't come to something like this'. [Earth Day event]
>
> During the habitat restoration, Miguel says, 'Mexicans don't do this kind of work for free. Why do you think we're the only Mexicans out here?'

As we have argued in earlier work (Tzou, Scalone, and Bell 2010), we see these comments as Miguel connecting the structured activity systems of community service to a larger narrative connecting Mexicans with a type of manual labour that they get paid for. In fact, while Miguel performs aspects of his Mexican identity in some situations, we also observed him resisting being positioned as a Mexican (or at least a Spanish speaker) in others:

> As one speaker tries to speak to the class in Spanish, Miguel says, 'I speak English too! You're just like my dentist, he tries to speak to me in Spanish too!'

While Miguel invokes his Mexican identity in moments of resistance to assert power from the margins of activity, having others impose it on him might be seen as an act of taking that power from him. With his comments about 'this kind of work', we can hypothesise that the associations he makes between 'being Mexican' and prevailing cultural narratives about what it means to be Mexican make him resist the fixed nature of that boundary. At the same time, however, he *lives* that identity fluidly, performing as a member of that group in certain learning situations.

Miguel's case is an illustration of the importance of understanding, through ethnography, the cognitive ecologies of youth in environmental education settings. Knowing Miguel's strong identity as a Mexican – and the ways in which he operationalises that identity – as well as the ways in which he views the connections between Mexicans, prevailing narratives about Mexicans, the connection between Mexicans and 'labour', and the connection between Mexicans and environmental issues, could help us understand better how and why he and other youth choose to engage in or resist the activity systems of environmental education.

Implications

We have documented the role of borders and acts of resistance that stem from the margins – both within environmental education and for youth as they perform their identities. We recognise the utility of categories in communicating about environmental hazards but our analysis also reveals how the construction of those categories positioned communities in ways that reified existing social and political power relationships. Youth do resist these constructions and, in fact, a closer look at their identity performance gives us further insight into identity formation within the places of environmental education in response to certain learning pathways and identities that are 'made available' (Brickhouse and Potter 2001) to youth. As we see in Figure 1, youths' individual narratives or cultural ecologies influence how they engage in activity systems such as learning activities in the boat trip, community service and so on. Simultaneously, prevailing cultural narratives – about safe and unsafe products, about how to engage in science, about what it means to be Mexican – shape and make available certain learning opportunities within activity systems. However, youth can be positioned – or position themselves either at the centre or more on the margins of the activity of environmental education. Therefore, to understand how and why youth engage in or resist activity systems, we need to examine both their personal ecologies (which are largely invisible and ignored in traditional educational settings) and prevailing norms and values that shape those activity systems.

Through our analysis, we also see moments of possibility where youth could have been engaged in challenging prevailing environmental narratives. The creation of this 'third space' (Gutiérrez, Rymes, and Larson 1995) can be a place for youth and environmental instructors to pose real challenges to narratives of fear and privilege, to create a 'transcendent script' of political and social transformation.

Finally, we see strong implications of this work to the design of learning environments in environmental education that take a social justice stance and take youths' cognitive ecologies into account. As we mentioned earlier, recent design-based research work a 6-8 week curriculum called *My Place in Puget Sound* (Tzou et al. 2009), we take a place-based approach to culturally relevant science teaching (cf., Lee 2007) in order to empower students to use ocean science concepts they learn about in the unit in the enactment and communication of community-based action. This action project is designed to address a locally relevant environmental issue, and students are asked to communicate the results of their project publicly in their community (handing out flyers, making presentations in their locally community centres, on a local website, etc.). We also utilise an ethnographic protocol called *self documentation* (Clark-Ibañez 2004) to have students' document environmental practices in their own so that teachers can better connect their everyday practices with science practices in the classroom. This curriculum has been enacted over two years by four teachers in a large urban school district in the Pacific Northwest and has reached approximately 500 students, over 50% of who are from groups historically underrepresented in STEM fields. Initial results indicate that through participation in this curriculum students gained a better understanding of the connection between their actions and the health of Puget Sound and strongly demonstrated a sense of personal relevance of the curriculum to their lives. There was also a significant increase in students' sense of responsibility to work for the health of the oceans and coastal areas (Bruner 2011).

Connecting ethnographic methods with curriculum design work, therefore, has the potential to explore the dialectical relationship between the local production of scientific knowledge and larger social and political narratives in environmental education. It also, however, broadens and complicates what it means to engage in scientific practices and further work needs to more deeply connect youths' everyday practices and identities with science and environmental science learning in classrooms and out of school settings.

References

Anzaldúa, G. 1999. *La Frontera/Borderlands: The new mestiza.* San Francisco: Aunt Lute Books.

Benson, C. 2003. The unthinkable boundaries of self: The role of negative emotional boundaries in the formation, maintenance, and transformation of identities. In *The self and others: Positioning individuals and groups in personal, political, and cultural contexts*, ed. R. Harré and F. Moghaddam, 61–84. Westport, CT: Praeger.

Blommaert, J., and D. Jie. 2010. *Ethnographic fieldwork: A beginner's guide.* Buffalo, NY: Multilingual Matters.

Bowers, C.A. 2001. Toward and eco-justice pedagogy. *Educational Studies* 32, no. 4: 401–16.

Brickhouse, N.W., and J.T. Potter. 2001. Young women's scientific identity formation in an urban context. *Journal of Research in Science Teaching* 38, no. 8: 965–80.

Bruner, A. 2011. My place in puget sound. Paper presented at the annual meeting of the American Fisheries Society, September, in Seattle, WA.

Bullard, R.D. 1990. *Dumping in Dixie: Race, class, and environmental quality.* Boulder, CO: Westview Press.

Calabrese Barton, A. 2001. Capitalism, critical pedagogy, and urban science education: An interview with Peter MacLaren. *Journal of Research in Science Teaching* 38, no. 8: 847–59.

Calabrese Barton, A. 2003. *Teaching science for social justice.* New York, NY: Teachers College Press.

Casey, E.S. 1996. How to get from space to place in a fairly short stretch of time: Phenomenological prolegomena. In *Senses of place*, ed. S. Feld and K.H. Basso, 13–52. Santa Fe, NM: School of American Research Press.

City Department of Planning and Development. 2006. *Characteristics of income (Census 2000 summary file 3).* [Details omitted to protect anonymity.]

Clark-Ibañez, M. 2004. Framing the social world with photo-elicitation interviews. *The Behavioural Scientist* 47, no. 12: 1507–27.

County Department of Health. 1997. *Leading causes of hospitalisation: Zip code. 1989–1993.* [Details omitted to protect anonymity.]

Emerson, R.M., R. Fretz, and L. Shaw. 1995. *Writing ethnographic interviews.* Chicago, IL: University of Chicago Press.

Furman, G.C., and D. Gruenewald. 2004. Expanding the landscape of social justice: A critical ecological analysis. *Educational Administration Quarterly* 4, no. 1: 47–76.

Giroux, H.A. 1992. *Border crossings: Cultural workers and the politics of education.* New York, NY: Routledge.

Goodall, H.L. Jr. 2000. *Writing the new ethnography.* New York, NY: AltaMira Press.

Grass, R., and J. Agyeman. 2002. Reorienting environmental education for environmental justice. http://www.ejrc.cau.edu/summit2/EnviroEducation/pdf (accessed October 25, 2008).

Gruenewald, D. 2003. Foundations of place: A multidisciplinary framework for place conscious education. *American Educational Research Journal* 40, no. 3: 619–36.

Gruenewald, D. 2004. A Foucauldian analysis of environmental education: Toward the socioecological challenge of the Earth Charter. *Curriculum Inquiry* 34, no. 1: 71–107.

Gutiérrez, K., B. Rymes, and J. Larson. 1995. Script, counterscript, and underlife in the classroom: James Brown versus *Brown v. Board of Education. Harvard Educational Review* 65, no. 3: 445–72.

Harré, R., and F. Moghaddam. 2003. Introduction: The self and others in traditional psychology and in positioning theory. In *The self and others: Positioning individuals and groups in personal, political, and cultural contexts*, ed. R. Harré and F. Moghaddam, 1–11. Westport, CT: Praeger.

Harré, R., and L. van Langenhove. 1999. *Positioning theory: Moral contexts of intentional action.* Malden MA: Blackwell Publishers.

Holland, D., W. Lachicotte, D. Skinner, and C. Cain. 1998. *Identity and agency in cultural worlds.* Cambridge: MA: Harvard University Press.

hooks, b. 1990. *Yearning: race, gender, and cultural politics.* Boston, MA: South End Press.

Jones, R. 2009. Categories, borders and boundaries. *Progress in Human Geography* 33, no. 3: 174–89.

Komunyakaa, J. 2002. *Dark Waters.* In *The colors of nature: culture, identity, and the natural world*, ed. A.H. Deming and L.E. Savoy, 98–112. Minneapolis, MN: Milkweed Editions.

Lee, C.D. 2007. *Culture, literacy, and learning: Taking bloom in the midst of the whirlwind.* New York, NY: Teachers College Press.

Lefebvre, H. 1991. *The production of space.* MA: Blackwell Publishing.

Lim, M., and A. Calabrese Barton. 2006. Science learning and a sense of place in an urban middle school. *Cultural Studies of Science Education* 1: 107–42.

MacLaren, P., and D. Houston. 2004. *Revolutionary ecologies: Ecosocialism and critical pedagogy. Educational Studies* 36, no. 1: 27–45.

Newman, D. 2006. The lines that continue to separate us: Borders in our 'borderless' world. *Progress in Human Geography* 30, no. 2: 143–61.

Newman, D., and A. Paasi. 1998. Fences and neighbours in the postmodern world: Boundary narratives in political geography. *Progress in human geography* 22, no. 2: 186–207.

Rodman, M.C. 1992. Empowering place: Multilocality and mulivocality. *American Anthropologist* 94, no. 3: 640–56.

Sobel, D. 1999. *Beyond ecophobia: Reclaiming the heart in nature education.* MA: The Orion Society.

Soja, E.W. 1980. The socio-spatial dialectic. *Annals of the Association of American Geographers* 70, no. 2: 207–25.

Solórzano, D.G., and O. Villalpando. 1998. Critical race theory, marginality, and the experience of students of color in higher education. In *Sociology of education: Emerging perspectives*, ed. C.A. Torres and T.R. Mitchell, 211–24. New York, NY: State University of New York Press.

Spradley, J. 1979. *The ethnographic interview.* New York, NY: Holt, Rinehart and Winston.

Strauss, A., and J. Corbin. 1998. *Basics of qualitative research: techniques and procedures for developing grounded theory.* Thousand Oaks, CA: Sage.

Tatum, B. 1997. *'Why are all the black kids sitting together in the cafeteria?' And other conversations about race.* New York, NY: Basic Books.

Thomas, M.E. 2005. 'I think it's just natural': The spatiality of racial segregation at a US high school. *Environment and Planning A* 37: 1233–48.

Tzou, C., A. Bruner, T. Clay, and G. Scalone. 2009. *My place in Puget Sound: Making connections between people, places, and marine environments.* Seattle, WA: Centre for Ocean Sciences Education Excellence–Ocean Learning Communities, University of Washington.

Tzou, C., G. Scalone, and P. Bell. 2010. The role of environmental narratives and social positioning in how place gets constructed for and by youth: Implications for environmental science education for social justice. *Equity and Excellence in Education* 43, no. 5: 105–19.

Veninga, C. 2009. Fitting in: The embodied politics of race in Seattle's desegregated schools. *Social and Cultural Geography* 10, no. 2: 107–29.

Wolcott, H.F. 1999. *Ethnography, a way of seeing*. New York, NY: AltaMira Rress.

Index